REFLECTIONS OF ARMENIAN IDENTITY IN HISTORY AND HISTORIOGRAPHY

Reflections of Armenian Identity in History and Historiography
Edited by H. Berberian & T. Daryaee

© Edited by H. Berberian & T. Daryaee 2018

H. Berberian & T. Daryaee are hereby identified as author of this work in accordance with Section 77 of the Copyright, Design and Patents Act 1988

Cover and Layout: Kourosh Beigpour | ISBN: 978-1-949743-01-2

All rights reserved. No part of this publication may be reproduced, stored in a retrieval system, or transmitted, in any form or by any means, electronic, mechanical, photocopying, recording or otherwise, without the prior permission of the pulishers.
This book is sold subject to the condition that it shall not, by way of trade or otherwise, be lent, resold, hired out or otherwise circulated without the publisher's prior consent in any form of binding or cover other than that in which it is published and without a similar condition including his condition being imposed on the subsequent purchaser.

REFLECTIONS OF ARMENIAN IDENTITY IN HISTORY AND HISTORIOGRAPHY

Edited by
H. Berberian & T. Daryaee

— 2018 —

Table of Contents

Preface
Introduction   9

Ancient   13

Gregory E. Areshian   15
Historical Dynamics of the Endogenous Armenian, i.e. *Hayots*, Identity: Some General Observations

Touraj Daryaee   37
The Fall of Urartu and the Rise of Armenia

Ani Honarchian   45
Of God and Letters: A Sociolinguistic Study on the Invention of the Armenian Alphabet in Late Antiquity

Khodadad Rezakhani   55
The Rebellion of Babak and the Historiography of the Southern Caucasus

Giusto Traina   71
Ambigua Gens? Methodological Problems in Ancient Armenian History

Early Modern   81

Sebouh David Aslanian   83
The "Great Schism" of 1773: Venice and the Founding of the Armenian Community in Trieste

S. Peter Cowe 133
The Armenian Oikoumene in the Sixteenth Century: Dark Age or Era of Transition?

Roman Smbatyan 161
Some Remarks on the Identity and Historical Role of Artsakhi Meliks in the Seventeenth and Eighteenth Centuries CE

Modern 169
Myrna Douzjian 171
Armenianness Reimagined in Atom Egoyan's *Ararat*

Shushan Karapetian 181
The Changing Role of Language in the Construction of Armenian Identity among the (American) Diaspora

Rubina Peroomian 189
Effects of the Genocide, Second Generation Voices

List of Contributors 205

PREFACE

This volume is the result of a conference held on the UCI campus in April of 2015. Armenian Identity Through the Ages: A Two-Day International Conference in Armenian Studies was made possible by the generous support of Vahe and Armine Meghrouni, Garo and Salpi Agopian, Charles and Mona Barsam, Alishan Halebian and Lydia Tutunjian, Vahag Hambarsumian and Seda Yaghoubian, Garbis and Zovak Karamardian, Krikor and Vehan Mahdessian, Viken and Arpi Melkonian, George and Hasmik Mooradian, Arek and Hanriette Tatevossian, Garo and Sylvie Tertzakian, Serge and Mona Tomassian, the Armenian General Benevolent Union, and the Orange County Armenian Professional Society. We thank UC Irvine Humanities and the Dr. Samuel M. Jordan Center for Persian Studies for their support and for making the conference possible.

We also recognize UCI's Armenian Studies Program, the Meghrouni Family Presidential Chair in Armenian Studies, the Maseeh Chair in Persian Studies and Culture, and the Samuel M. Jordan Center for Persian Studies for their efforts to actualize the present volume.

The purpose of this international conference was to explore various aspects of Armenian identity from the remote past to the present. Some of the papers that appear in this collection stay true to their original presentations while others have been dramatically altered, even in subject in one case. We would like to thank Kourosh Beigpour for the design and making the book camera-ready, and we are grateful to Virginie Rey for her diligent copyediting.

Houri Berberian	Touraj Daryaee
Meghrouni Family Presidential Chair in Armenian Studies	Maseeh Chair in Persian Studies & Culture

INTRODUCTION

Houri Berberian and Touraj Daryaee

According to WorldCat, there are almost 3,000 works that bear the terms "Armenia/n" and "identity" in their titles, of which over 900 are books. One might ask, why the need for another book on Armenian identity? We hope this eclectic collection spanning a period between the ancient and the present will help demonstrate the continuing relevance of the broad and complex concept and study of identity. The chapters express different interpretations and contexts of Armenian identity and demonstrate the multiple ways of approaching the oft-used term and concept of identity. Thus, the collection as a whole is also a reflection of historiographical developments and directions. The volume, however, is neither comprehensive nor homogeneous. While the overarching theme of Armenian identity, history, and historiography are at the forefront, the chapters have wide-ranging foci and, as a whole, cover a period of some three thousand years. The volume is chronologically organized into three sections – ancient, early modern, and modern – and then alphabetically within each section.

Ancient

The chapters in this section span several centuries and cover a wide variety of topics all centered on identity: from the long durée view of Armenian identity to the significance of Armenian place names, Armenian alphabet, regional interactions, and finally comparative perspectives,

Gregory Areshian's paper provides an overview of the issue of Armenian identity from antiquity to the present. He begins his work with the discussion of the idea of "ethnicity" itself. He criticizes previous works on the subject of Armenian identity pointing out that those who have engaged in the endeavor have not explored a broad historical longue durée perspective and, therefore, have distorted our perception of Armenian identity. Areshian then begins his survey from the first millennium BCE with Urartu and the Achaemenid view of Armenia/Armina and Armenians and moves forth to late antiquity to the early Islamic period where by 701 CE it is called the province of Armina/Arminia. Touraj Daryaee's discussion also takes us to Armina. Daryaee tackles the two terms associated with Armenia, namely Uraštu and Armina, and suggests that the important moment for the shifting of identity and change of power between the Urartians and the Armenians came in the sixth century BCE, when the Armenians sided with the Persians against the Urartians who had rebelled against the Persian king, Darius I. Hence, Armina and Armenians became the designation for the territory and the people in the Caucasus.

Ani Honarchian's essay addresses the invention and origin of the Armenian alphabet through Koriwn, the pupil of Maštocʻ, who chronicles his saga in search of a perfect alphabet, created at the court of the Arsacids. She suggests that regardless of what the origin of the alphabet, it was presented as a sacred, awe-inspiring, and distinctive gift from God to Armenians.

Khodadad Rezakhani focuses on Babak Khorramdīn, a ninth century rebel in the province of Azarbījān, and the regional interactions between the Dailam and the Caspian Coastal region with Arminiyah/Armenia. Rezakhani points out that Babak has been seen within either the Iranian or Islamic sphere of history, and his regional significance has not been considered by historians of Armenia and the region. He argues, instead, that Babak's rebellion must primarily be seen within the context of Armenian and Dailamite history and the negotiations of relationships between the local rulers of each region and the central Islamic authority.

Giusto Traina explores the importance of Armenia in the ancient world and provides a comparative study of smaller kingdoms/states. He finds Armenia's designation as a "buffer state" inadequate, seeing it instead as an independent kingdom for about six-and-a-half centuries, when it held a strategic position in antiquity. However, the image of Armenia appears to have been fluid without a firm territory, thus connecting it much more meaningfully to both Asia and the Mediterranean.

Early Modern

The chapters in this section explore early modern Armenian identity from very different angles but all to one degree or another challenge received tradition. While Sebouh Aslanian reassesses the Mkhitʻarists' historic division in the eighteenth century and Peter Cowe reevaluates sixteenth-century Armenian literary production, Roman Smbatyan considers the multi-level perception of Artsakhi Melik identity.

Sebouh Aslanian's chapter addresses the 1773 schism in the Armenian Catholic Mkhitʻarist congregation, which played a fundamental role in the preservation of the Armenian language and literature. Aslanian makes extensive use of heretofore unconsulted documents from state archives in Venice, Trieste, Rome, and Vienna, among other evidence, to argue against the grain that what lay behind the schism which tore apart the congregation was neither theological nor sectarian; rather, it was the consequence of quarrels over constitutionalism and representative monastic governance as well as anxieties over mismanagement of funds. Aslanian traces the history of the order and its founder, Abbot Mkhitʻar, the circumstances leading to the rupture, and as a consequence, the creation and history of the Armenian community in Trieste (and then Vienna). The analysis ends with a discussion of the reasons behind the schism, concluding – based on new evidence – that the origins of the conflict that separated the congregation into two branches lay in disagreements over authority and worries over mishandling of finances. Aslanian's contribution goes to the core of the Armenian congregation's identity by challenging its representation (self and other) and providing an alternative view of the split that marked its history, creating two "often rival, bitterly factional orders."

Like Aslanian, Peter Cowe challenges historiographical tradition. In this case, Cowe questions the designation of the sixteenth century in the Armenian case as a Dark Age, characterized by an absence in cultural, literary, and historical production, which ignores chroniclers like Grigor Daranałc'I, Samuēl Anec'i, Yovhannēs Arčišec'I, and Yovhanisik Carec'i. Cowe instead argues for a reassessment of sources as well as a reconceptualization of the period. An examination of contemporary literary works such as martyrologies, translations, and fictional works lead Cowe to conclude that sixteenth-century literary achievements corresponded to sociopolitical and commercial developments. By reminding us of connections between different facets of Armenian society, developments on the Plateau and diaspora centers, and the links of communities among wider networks of exchange, Cowe provides a different perspective on Armenian identity and historiography. He contends that these developments denote Armenian agency on a global dimension – "far from Armenians entering into hibernation or living an inward-looking ghettolike existence."

Unlike previous scholarship which has focused on the military, political, and cultural roles of Artsakhi Meliks, Roman Smbatyan explores the identity of Artsakhi Meliks, who were in the vanguard of medieval and early campaigns for liberation from foreign and local aggression and for Armenian statehood. Through an examination of Armenian, Russian, and Persian primary sources and archival materials, Smbatyan analyzes Artsakhi Melik identity from three angles: first, self-ascription, that is, how the Meliks perceived themselves; second, Armenian Diasporan view; and third, foreign, that is, Russian officials' and Persian historians' opinions. He argues that an evaluation of all three levels of perception in the early modern period points to a congruity in outlook that identified the Artsakhi Meliks as key guardians of Armenian identity and statehood.

Modern

This section sets out to complicate Armenian identity/identities and their narrative further through the lens of film, language, and literature. Myrna Douzjian and Rubina Peroomian both look at the way in which the Armenian Genocide has shaped diasporan Armenian identity although their methodologies and approaches are quite different while Shushan Karapetian explores the complex relationship between Armenian identity and language.

Myrna Douzjian argues against the simplistic assertion that the Armenian Diasporan collective consciousness has in its cultural (including cinema) and literary production perceived the Catastrophe, that is, the Armenian Genocide, in the same way. By privileging "proof narratives" whose aim is to prove Genocide and refute denial, the complexities of survivor memories have fallen by the wayside, thus uniformizing Armenian identity. Through a rereading of Atom Egoyan's *Ararat* (2002) and a review of the unsympathetic reception of the film, Douzjian argues that the censure of the film from various corners was the consequence of its subversion of the proof narrative and binaries of self/other, truth/lie, and history/fiction. *Ararat* does not concern itself with retelling a factually grounded story of the Catastrophe. Douzjian reminds us that instead, the film focuses on the agency of individuals and the role of harrowing narratives in constructing identity, thus asserting that identity is "a process that relies on cultural production."

Shushan Karapetian's chapter pursues another side of the coin of identity construction. Karapetian notes the intimate connection between language and identity, as speakers negotiate language communities as well as their own identities. Through an analysis of interviews and by tracing the development of the Armenian language through generations of Armenian-Americans, Karapetian argues that the language as a form of communication has decreased (as apparent in the first generation) and, instead, its symbolic value privileged (as in second and third generations), thus in some way limiting or severing the dependence of identity on language knowledge and use. This "symbolic 'Armenianness'" accompanies, however, an internalized sense of moral obligation to safeguard against assimilation. The consequence is the perception that Armenian has little, if any, real-world utility or benefit and, therefore, is relegated to the emotional and symbolic domain. However, Karapetian demonstrates, this imbalance between low or no language proficiency and increased and moral responsibility has led to high levels of anxiety and shame as younger generations feel guilt about being unable to transmit Armenian identity to their children through language and consequently culture and history.

The last chapter of this section by Rubina Peroomian returns to the relationship between the Armenian Genocide and its role in shaping diasporan Armenian identity. Peroomian studies the writings of second-generation Armenian writers and looks at the psychological impact and transmission of the trauma of survivor parents on children. She delves into the post-genocidal environment, parent-children relationships (for example, whether parents discussed their experiences), to what extent were the second-generation Armenian writers detached from their own family and ethnic history, and so forth. She argues that parent-children relationships were the inspiration or provocation that led to literary responses, whether parents were able to share their traumatic experiences with their children or not. According to Peroomian, these children were moved to pen their expression as a way to confront the Armenian Genocide, and their writings reflect "the nation's collective psyche." Here, we see the significance of the links between genocide, the conceptualization of identity, and literary production.

Note: Transliteration of names and terms from Armenian has been left up to each author's preference.

ANCIENT

HISTORICAL DYNAMICS OF THE ENDOGENOUS ARMENIAN, I.E. *HAYOTS*, IDENTITY: SOME GENERAL OBSERVATIONS

Gregory E. Areshian

The academic and societal interests in the topic of Armenian identity have a recent history. They originated in the course of the last several decades, not lagging too far behind the emergence of the current theoretical framework of studies in ethnic identity, which occurred in the Anglophone world towards the second half of the 1950s. With regard to the latter, it is remarkable that, during that decade, the term "ethnic identification" was still semantically used in anthropological narratives as a term defining research procedures identifying cultural frontiers, contact zones, or specific cultural features with ethnicities within the framework of ethnic history,[1] and studies of archaeological cultures. But one finds, almost at the same time, the beginnings of current conceptualization of ethnic identity as "ethnic identification" by sociologists,

> "Ethnic identification" refers to a person's use of racial, national, or religious terms to identify himself, and, thereby, to relate himself to others. "Ethnic orientation" refers to those features of a person's feelings and actions towards others, which are a function of the ethnic category by which he identifies them. Ethnic identification and orientation are seen as two aspects of a single behavioral complex.[2]

Thus, a paradigmatic shift from the identification of ethnic features to the investigation of ethnic identity as something having an ontological status took place around that time. Yet the roots of this current conceptualization are much deeper. One can find them in Johann Gottlieb Fichte's philosophy of the patriotic-cosmopolitan national self-awareness,[3] and

---

[1] E.g. Gene Weltfish, "The Question of Ethnic Identity, an Ethnohistorical Approach," *Ethnohistory* 6 (1959): 321-46.

[2] Daniel Glaser, "Dynamics of Ethnic Identification," *American Sociological Review* 23 (1958): 31.

[3] Hans Kohn, "The Paradox of Fichte's Nationalism," *Journal of the History of Ideas* 10, no. 3 (1949): 325-27.

follow them through the dark and bright vagaries of nationalistic, patriotic, and cosmopolitan consciousness, at the end finding them adjoining, most of the time indirectly, recent theories of social identity developed by social psychologists, among which the theory of Henry Tajfel[4] keeps attracting a major following.[5] The studies of ethno-cultural and ethno-political identities, as classes of social identity, have a promising future and, in the meantime, they gradually subsume, partially or entirely, different aspects of earlier concepts and research areas, such as the German "peoples' spirit," or *Völksgeist* (Herder/Humboldt/Hegel et al.), Wilhelm von Humboldt's *Nationalcharakter* "national character," *Völkerpsychologie* "peoples' psychology," *Weltansicht* "world view" derived from and embodied in a natural language, Ernest Renan's *esprit de la nation* "national mindset," — "a spiritual principle resulting from the profound complexities of history,"[6] or, again in French, "collective mentality," and, finally, "lifeway" (a recently revitalized term, which, however, appears in the Anglo-Saxon *Exeter Book*, Gutlac A, according to the *Oxford Living English Dictionary*, 2017 Online Edition)/"way of life," or "folkways," "national forms of life," and "national belonging."[7]

---

[4] Henry Tajfel, "Social Identity and Intergroup Behavior," *Social Science Information* 13, no. 2 (1974): 65-93; Henry Tajfel, *Human Groups and Social Categories: Studies in Social Psychology* (Cambridge, UK: Cambridge University Press, 1981); Henry Tajfel, "Social Psychology of Intergroup Relations," *Annual Review of Psychology* 33 (1982): 1-39.

[5] Rupert Brown and Dora Capozza, eds., *Social Identity Processes: Trends in Theory and Research* (London: SAGE Publications, 2000); B. Curtis Eaton, Mukesh Eswaran, and Robert J. Oxoby, "'Us' and 'Them': The Origins of Identity and its Economic Implications," *The Canadian Journal of Economics* 44, no. 3 (2011): 721; stressing the evolutionary underpinnings of social identity as a mechanism facilitating survival, Michael A. Hogg, "Social Identity Theory," in *Understanding Peace and Conflict Through Social Identity Theory*, eds. S. McKeown, R. Haji, and N. Ferguson (Berlin-Heidelberg-New York: Springer Verlag, 2016), 3-17.

[6] Ernest Renan, "Qu'est-ce qu'une nation?," *Bulletin de l'Association Scientifique de France* (26 March 1882).

[7] William Graham Sumner, *Folkways: A Study of the Sociological Importance of Usages, Manners, Mores, and Morals* (Boston, Mass.: Ginn & Co, 1907); Kurt Danziger, "Origins and Basic Principles of Wundt's *Völkerpsychologie*," *British Journal of Social Psychology* 22 (1983): 303-13; Matti Bunzl, "Franz Boas and the Humboldtian Tradition: From *Volksgeist* and *Nationalcharakter* to an Anthropological Concept of Culture," in *Volksgeist as Method and Ethic: Essays on Boasian Ethnography and the German Anthropological Tradition*, ed. George W. Stocking, Jr. (Madison: University of Wisconsin Press, 1996), 20-33; Avishai Margalit, "The Moral Psychology of Nationalism," in *The Morality of Nationalism*, eds. Robert McKim and Jeff McMahan (Oxford: Oxford University Press, 1997), 82-5; Nenad Miscevic, "Nationalism," *The Stanford Encyclopedia of Philosophy*, ed. N. Edward Zalta (Winter 2014 Edition), https://plato.stanford.edu/archives/win2014/entries/nationalism/

In the course of the last decade and a half, Identity Studies, integrating individual and social identities, gradually became recognized as an interdisciplinary field of the Social Sciences and the Humanities, with all the deriving theoretical and practical implications.[8] And if we look at the whole span of broadest approaches toward ethno-cultural identity, from the impact of the natural environment on the shaping of identity in Fernand Braudel's *Histoire Profonde*, "deep history" of the French identity,[9] to Richard Nisbett's[10] insights concerning the ethno-geographic and civilizational specificities of thought, it would be hard to avoid the conclusion that cultural identity is an extremely complex amalgam of collective and individual perceptions, assessments, and related spiritual and tangible expressions answering foundational questions of orientations and values at an ethno-cultural (including religious), ethno-political (from tribal and communal to national), and civilizational levels of sociocultural organization of humankind. Therefore, the spectrum of social sciences and the humanities, from psychology, linguistics, philosophy and history to anthropology, sociology, literary criticism, folklore, archaeology, art and religion history, would have to join their efforts in developing Identity Studies.

While general transdisciplinary theories and methodologies of Identity Studies are still emerging, some theoretical considerations are due in order to further the case studies in ethno-cultural and ethno-political identities.

First, the phenomenal contradiction between the instability, fluidity, and transitive character of personal ethno-cultural identities demonstrated in psychological experiments and through many ethnographic-anthropological observations[11] and the long-term persistence of cultural identities on a personal and, especially, group levels documented in sociological surveys and historical research requires further exploration and understanding.[12]

---

[8] James Côté, "Identity Studies: How Close Are We to Developing a Social Science of Identity? – An Appraisal of the Field," *Identity: An International Journal of Theory and Research* 6, no. 1 (2006): 3-25 ; Joel R. Sneed, Seth J. Schwartz, and William E. Cross Jr., "A Multicultural Critique of Identity Status Theory and Research: A Call for Integration," *Identity: An International Journal of Theory and Research* 6 (2006): 61-84.

[9] Fernand Braudel, *The Identity of France*, vol. 1, *History and Environment*, trans. Siân Reynolds (London: Collins, 1988). Fernand Braudel, *The Identity of France*, vol. 2. *People and Production*, trans. Siân Reynolds. (London: Collins, 1990).

[10] Richard E. Nisbett, *The Geography of Thought: How Asians and Westerners Think Differently... and Why* (New York: Free Press, 2003).

[11] Daniel Glaser, "Dynamics of Ethnic Identification," 35-7; Jenny K. Phillips, *Symbol, Myth, and Rhetoric: The Politics of Culture in an Armenian-American Population*, in Immigrant Communities and Ethnic Minorities in the United States and Canada, vol. 23. (New York: AMS Press, 1989): 6-9; Alexander Kuo and Yotam Margalit, "Measuring Individual Identity: Experimental Evidence," *Comparative Politics* 44, no. 4 (2012): 459-79; Eric T. Olson, "Personal Identity", *The Stanford Encyclopedia of Philosophy*, ed. Edward N. Zalta (Spring Edition, 2016), https://plato.stanford.edu/archives/spr2016/entries/identity-personal/

[12] Richard D. Alba and Mitchell B. Chamlin, "A Preliminary Examination of Ethnic Identification Among Whites," *American Sociological Review* 48, no. 2 (1983): 246-247; Eric T. Olson, "Personal Identity."

At first sight one may give preference to psychologists' conclusions considering their more rigorous methods of experimental investigation. But a deeper analysis of these studies reveals the fact that at least some of them did not take sufficiently into consideration the differences between specific historical contexts of ethno-cultural perceptions subjected to investigations.

Another key to the interpretation of this contradiction could be sought in Wilhelm Wundt's distinction between the two levels of social action and communication – *Triebbewegung*, "drive action," and *Willkürbewegung,* "discretionary action." Confirmed both experimentally and historically, this fundamental distinction,

> ...implies two levels of human social interaction. On the first level interaction is not governed by individual attentions but by a group process; on the level of discretionary action, however, individuals do interact *as individuals* with their own independent goals and intentions. The distinction is an analytic one; in the adult person both levels of social behavior coexist and interact with one another... The adult individual therefore remains part of collective processes, which occur without his intentional contribution.[13]

But, unlike Wundt, one may argue, on the grounds of historicity, that it is the discretionary actions of individuals that cumulatively impact the group process. That is another expression of the dialectic interaction between tradition and innovation, which forms the foundation of the sociocultural evolution. Social identity and its transformations are an essential part of that evolutionary process and are subject to its regularities and patterning.

Second, since social identities should be studied from the perspective of interaction between persistence and transformation, one cannot accept Herder's assertion that *Volk*, understood by him as a "primary cultural community,"[14] is the principal creator and carrier of ethno-cultural identity opposed to the anti-ethnic cosmopolitanism of the 18th century ruling quasi-feudal elites, which later was reduced by Tönnies[15] and Weber[16] to the opposition between *Gemeinschaft* (community based on personalized ties, dependences, kinship, and affections) and *Gesellshaft* (society based on formal relations, impersonal values, and rational agreement). German history has demonstrated on its own that the ethno-cultural traditionalism of the "primary cultural community" was proactively reshaped in the course of the nineteenth century by the political and intellectual elites into the German ethno-national identity and used for political purposes, which played an important role in the victory of Bismarck's German Empire. Such a metamorphosis of ethno-cultural into ethno-

---

[13] Kurt Danziger, "Origins and Basic Principles of Wundt's *Völkerpsychologie*," 309.

[14] Frederick M. Barnard, *Herder's Social and Political Thought: From Enlightenment to Nationalism* (Oxford: Clarendon Press, 1965).

[15] Ferdinand Tönnies, *Gemeinschaft und Gesellschaft* (Leipzig: Fues's Verlag, 1887).

[16] Max Weber, *Economy and Society*, eds. Guenter Roth and Claus Wittich (Berkeley and Los Angeles: University of California Press, 1968), 40-3.

political (including ethno-national) social identities driven by innovating social leaderships is observable throughout human history. These metamorphoses may be considered as main pivots in the ethnic histories of peoples and, as such, they deserve further exploration through particular case studies.

Third, an exceptionally important aspect of identity of an established social group is the symbolization of that identity by verbal or non-verbal means, ranging from family habits and stories to monumental architecture, and from entire languages as ethno-social markers to mascots, coats of arms, and logos conveying messages of identity.[17] Therefore, comprehensive investigations of symbolic systems of identity adhered to and produced by social (including ethno-cultural and ethno-political) groups should be of primary interest for Identity Studies. Understanding symbolization makes social identities intelligible.[18] One may add to this the observation by E. K. Francis that, "...it will be quite legitimate to state that some **concrete social group is an ethnic group to a lesser or greater degree.**"[19]

Fourth, the relationship between the second (persistence, variability, and change in ethno-cultural and ethno-political identities) and the third (symbolization of social identities) aspects could be elucidated through the study of social and cultural production, innovation, intensification, expansion, and contraction, understood in a broader perspective than Pierre Bourdieu's theory of fields of cultural production, habitus, and praxis interpreted through the prism of power relations and capitalist economy.[20] The robustness and vitality of social identities and their decline or disintegration could and should be explored and interpreted in direct and indirect correlations with cultural production.

Fifth, since the theoretical approaches toward social Identity Studies have to be framed by multiple correlated sets of oppositions and comparisons, such as I/you, I/she, I/others, I/they, you/she, you/others, you/they, we/they, we/others, etc., the aforementioned dichotomy of "ethnic identification" vs. "ethnic orientation" unjustifiably narrows the field of investigation of collective social identities, which should be construed at least as triadic and include (a) a group's perception of itself (which we will call ***endogenous perception of group identity***), (b) ***a group's perception of other groups***, (c) a group perception **by** other groups (which we will call ***exogenously constructed characterization***), with adequate internal analytical subdivisions within these three interacting perspectives. So, for example, the expression/projection of desirable political identity of a group and its leaders would constitute a part of (a).

---

[17] For national identity and its symbolization, see Anthony D. Smith, *National Identity* (Reno: University of Nevada Press, 1991), 74-9.

[18] With regard to this statement, the author of this paper acknowledges an influence from Pierre Bourdieu's ideas.

[19] E. K. Francis, "The Nature of the Ethnic Group," *American Journal of Sociology* 52, no. 5 (1947): 400.

[20] David Rasmussen, "Praxis and Social Theory (Review of *Outline of a Theory of Praxis*)," *Human Studies* 4, no. 3 (1981): 273-78; David Gartman, "Bourdieu's Theory of Cultural Change: Explication, Application, Critique," *Sociological Theory* 20, no. 2 (2002): 255-77.

Sixth, the three perspectives mentioned in the preceding, i.e. fifth observation, need to be explored on the intersections between the sociocultural paradigms/lifeways of the primary cultural communities and the agencies of proactive leaderships introducing changes into cultural production.

A simultaneous development of these six research directions certainly will advance Identity Studies in general and the investigations of Armenian identity in particular, the latter still being at a very early stage of their development. Indeed, if one searches Google Scholar for the words "Armenian identity," she would find only a handful of cases during the 1950s when these words are found together, but without any elaboration, analysis, or interpretation.[21] Almost in all of the 320 academic works published during that decade in English, which separately use the words "Armenian" and "identity" in a single text, "identity" is used as the synonym of the research process of "identification," most of the time related to the establishment of authorship of different works, i.e. the very concept of Armenian ethno-cultural identity didn't exist at that time, within the framework of academic research. The situation did not change during the 1960s, and it was only in the 1970's that the first scholarly works particularly addressing some aspects of modern Armenian identity were written.[22]

Different societal factors stimulated an increase in the academic interest toward Armenian identity from 1980s onwards. It was the awakening of a deep concern amid the Armenian diaspora regarding the future of its ethno-cultural self-preservation and a broader intellectual imperative to study and interpret Armenian identity stemming from the strengthening of patriotism and the rise of nationalism, the Karabagh movement, and the public demand for the recognition of the Armenian Genocide that have become the societal forces driving the research in Armenian identity. As it happens most of the time (fortunately not always), social movement preceded the deep scholarly study of the problem. So, the pan-Armenian movement devoted to the recognition of the Genocide, which grew and strengthened in earnest in the Republic of Armenia and the Armenian diaspora since 1965, preceded the rise of academic interest in Armenian identity by nearly two decades.

---

[21] John P. Cavarnos, "Review of: Paul Peeters *Le tréfonds oriental de l'hagiographie byzantine,*" *Speculum* 26, no. 4 (1951): 735-38.

[22] e.g. Jenny K. Phillips, *Symbol, Myth, and Rhetoric.* (Data collected in 1974-1977, concerning the character of Armenian ethno-cultural identity driven by local diasporic politics in the Armenian communities of the USA).

The number of publications partially discussing or completely focused on the studies of Armenian identity at least tripled since the year 2000. Three landmark books devoted to (a) a comprehensive anthropological exploration of the broadest spectrum of manifestations of current Armenian identity,[23] (b) a sociological study of the Armenian diaspora in the USA,[24] and (c) to the perception of Armenians by non-Armenians,[25] the latter concerning what I referred to above as the exogenously constructed characterization, have been published since the emergence of a consistent academic interest in this subject. Some more particular topics, such as the contemporary iconography of Armenian identity,[26] the revitalization and partial re-creation of modern Armenian identity by ethno-cultural and political movements in the 19th century,[27] and expressions of Armenian identity in modern folk culture[28] have been explored in-depth.

But, notwithstanding this remarkable progress, four essential directions in studies of Armenian identity have not been developed yet, even on a minimally desirable scale.

First, as of today, most of these studies did not adequately grasp the importance and, consequently, didn't internalize the major advancements in theories of social identity.

Second, the preoccupation with problems of Armenian identity in our contemporary world without an exploration of those in a broad historical perspective of the *longue durée*, leads to a distortion of the problematique, perpetuates a presentist bias, and undermines the exploration and interpretation of historical causality in these studies. It is true that, with the emergence of the whole field of studies in Armenian identity, the origins of modern Armenian identity have been sought. The realization of the importance of the Armenian Apostolic Gregorian Church in the sustenance of that identity, during the Medieval and Modern times, has naturally brought into the research spotlight the question of the earlier role of the Church in the shaping of that identity. Consequently, the development of the specific character of the Armenian Church and its religious and sociopolitical positioning vis-à-vis other religions, and sociopolitical institutions and structures, in its different manifestations and in connection with the Armenian identity, attracted much attention, mostly within the chronological

---

[23] Levon Abrahamian, *Armenian Identity in a Changing World,* Armenian Studies Series, no. 8 (Costa Mesa: Mazda Publishers, 2006).

[24] Anny P. Bakalian, *Armenian-Americans: from Being to Feeling Armenian* (New Brunswick: Transaction Publishers, 1993).

[25] Richard G. Hovannisian, ed., *The Armenian Image in History and Literature* (Malibu: Undena Publications, 1981).

[26] Harutyun Marutyan, *Iconography of Armenian Identity,* vol. 1, *The Memory of Genocide and the Karabagh Movement* (Yerevan: Gitut'yun Publishing House, 2009).

[27] Lisa Khachaturian, *Cultivating Nationhood in Imperial Russia: The Periodical Press and the Formation of a Modern Armenian Identity* (New Brunswick: Transaction Publishers, 2009).

[28] Levon Abrahamian and Nancy Sweezy, eds., *Armenian Folk Arts, Culture, and Identity* (Bloomington: Indiana University Press, 2001).

framework of the sixth, seventh, and eight centuries CE.[29] Other interdisciplinary studies explored external political factors in the construction of Armenian identity[30] and the roots of several important ethno-cultural symbols.[31] Finally, a valiant and useful attempt has been made to summarize different aspects (including the historical) of studies in Armenian identity and present those in a collection of essays,[32] despite the current gaps between different empirical studies and deficiencies in theoretical and methodological approaches. All things considered, one may conclude that all that important work still represents an early stage in historical exploration of Armenian identity, being limited most of the time to the study of the endogenous perception of Armenian ethnic identity. The historical hierarchies and dynamics of that identity still await further investigation.

Third, constituting an integral part of researches in Armenian identity, systematic studies of the Armenians' perceptions of other ethno-cultural and ethno-political groups are almost non-existent, especially from a historical perspective, even though the exploration of extant factual material looks quite promising.[33]

Fourth, the dearth of studies of the ethnic perceptions of Armenians by other ethno-cultural and political groups (i.e., the exogenously constructed characterization) leads to a substantial confusion and misinterpretation of the Armenian past and present.

---

[29] Nina G. Garsoïan and Jean-Pierre Mahé, *Des Parthes au Califat: Quatre leçons sur la formation de l'identité arménienne*. Monographies du Centre de Recherche d'Histoire et de Civilisation de Byzance, vol. 10 (Paris: Editions de Boccard, 1997); Boghos Levon Zekiyan, "Quelques réflexions préliminaires sur l'identité chrétienne de l'Arménie: L'universalité de la parole et son incarnation dans la vie de l'ethnos," *Connaissance des Pères de l'Eglise* 81 (2001): 21-37; Anne Elizabeth Redgate, "Myth and Reality: Armenian Identity in the Early Middle Ages," *National Identities* 9, no. 4 (2007): 281-306; Theo Maarten Van Lint, "The Formation of Armenian Identity in the First Millennium," in *Religious Origins of Nations? The Christian Communities of the Middle East*, ed. B.T.H. Romeny (Leiden: Brill, 2009), 251-78; Nina G. Garsoïan, *Interregnum. Introduction to a tudy on the Formation of Armenian identity (ca. 600–750)* (Louvain: Aedibus Peeters, 2012); Tara L. Andrews, "Identity, Philosophy, and the Problem of Armenian History in the Sixth Century," in *History and Identity in the Late Antique Near East*, ed. Philip Wood (Oxford-New York: Oxford University Press, 2013), 29-42.

[30] Gregory E. Areshian, "Sasanian Imperialism and the Shaping of Armenian Identity: Interdisciplinary Verification and Ambivalence of Empire-Nation Relationship," in *Empires and Diversity: On the Crossroads of Archaeology, History, and Anthropology. Ideas, Debates, and Perspectives*, ed. G. E. Areshian (Los Angeles: Cotsen Institute of Archaeology Press, University of California, 2013), 146-63.

[31] Hamlet L. Petrosyan, "Symbols of Armenian Identity," in *Armenian Folk Arts, Culture, and Identity*, eds. L. Abrahamian and N. Sweezy (Bloomington: Indiana University Press, 2001), 25-71.

[32] Edmund Herzig and Marina Kurkchiyan, eds., *The Armenians: Past and Present in the Making of National Identity* (London & New York: Routledge, 2005).

[33] Petra Kohoutková, *Image de l'Autre dans les sources arménienne du XVIe au XVIIIe siècles: Naissance et évolution d'un stéréotype ethnique*. Ph.D. Dissertation, University Montpellier 3, France 2008.

A salient example of the confusion resulting from the absence of a clear-cut distinction between the endogenous perception of Armenian identity and its exogenously constructed characterization is the insufficient and often inadequate interpretation of historically attested semantic discrepancy between the endonyms *hay, hayots, Hayk', Hayastan* and the exonyms "an Armenian," "Armenian," "Armenia." Under the historical circumstances of modern times, it is assumed as self-evident that *hay* is understood and translated as "an Armenian," and *Hayastan* as "Armenia." It can be argued, beyond reasonable doubt, that these two sets of endonyms and exonyms historically have been closely related, forming different aspects of characterization of identity. But it is also easy to demonstrate that, most of the time, they were not signifiers of the same signified, and projecting todays' indiscriminate synonymization of these two sets of signifiers with one another into the past is presentist and thereby erroneous.

Going back all the way to the end of the sixth century BCE, one finds in the Bisotun (Behistun) inscription of the Achaemenid Darius I, carved within a couple of years after 521/519 BCE, references to *Armina*/Armenia that is described as an enormous country *dahyu*, located between, *Mâda*/Media (today's central Western Iran) and *Katpatuka*/Cappadocia (today's central Turkey), among the 23 countries that formed the Achaemenid Empire.[34] In the south, *Armina* bordered on Assyria, whereas, in the north, it was reaching the Black Sea.[35] In no times indigenous sources written in Armenian language had included such a vast territory within the limits of *Hayk'/Hayastan*. Those sources had located to the east/southeast of *Hayastan* the country of *Atrpatakan*, i.e., the Kingdom of Media-Atropatene of the Hellenistic period, describing its place between *Hayk'/Hayastan* and *Mâda/Marastan/marats ashkharh*/Media.

Darius carried out an administrative reorganization of the empire, first of all, appointing new satraps from the ranks of his closest supporters and relatives to govern those *dahyu*-satrapies. We do not know exactly how sweeping and transformative were those administrative reforms, yet Herodotus, despite all the criticism levied against him for his hellenocentrism,[36] captured, most likely correctly, the essence of Darius' reform when he wrote, "...he [Darius] instructed each people to pay him tribute, consolidating neighboring peoples and distributing outlying peoples among different provinces, passing over those adjoining."[37] Darius' satrapies became complex, semi-autonomous, polities that included both ethno-cultural *Volksgemeinschaft* i.e., "a unity of a people communities" ("national community" in modern times, a term originating from Tönnies' works) under the

---

[34] Roland G. Kent, *Old Persian: Grammar, Texts, Lexicon* (New Haven: American Oriental Society, 1950), 119, §6. 1.12-7).

[35] Bruno Jacobs, "Achaemenid Satrapies," in *Encyclopædia Iranica*, ed. Ehsan Yarshater (New York: Columbia University Press, 2011) 143-86,
http://www.iranicaonline.org/articles/achaemenid-satrapies

[36] O. Kimball Armayor, "Herodotus' Catalogues of the Persian Empire in the Light of the Monuments and the Greek Literary Tradition," *Transactions of the American Philological Association* 108 (1978): 1-9.

[37] Herodotus of Halicarnassus, 1920/2010, *The Histories*, trans. E.D. Godley (E-Book: Pax Librorum), III, 89.

sociopolitical governance of satrapal administration and ethno-political units of lower rank (such as tribes or tribal unions). It is important to stress that the sociopolitical, economic, and ethno-cultural diversity of satrapies were predetermined, to a large degree, by the historical pathways of a particular *dahyu*, shaped before their inclusion in the Achaemenid Empire. However, it is also important to underscore that the Achaemenid propaganda disseminated through imperial monumental art, does not reflect the internal diversity of satrapies, creating and projecting instead an iconic ethno-cultural/ethnographic image of representatives of a titular people perceived as symbolic of a satrapy. It is quite remarkable that Xenophon's account of the retreat of the corps of Greek mercenaries from Mesopotamia, through the *dahyu*-satrapy of Armenia to the Black Sea in 401-400 BCE, recorded approximately six generations after Darius' administrative reform, confirms in essence Herodotus' aforementioned characterization of the organization of an Achaemenid satrapy (Xenophon 1922).[38] According to Xenophon's narrative, three distinctively different kinds of political, ethno-political, and ethno-cultural subsystems were integrated within a single satrapy: (1) the satrap's bureaucratic apparatus and, partially, the military force dominating the administrative-geographic territory of the satrapy of Armenia that included many different ethnicities; (2) the semi-independent ethno-political territories of the *Kardoukhoi, Khaldaioi*, and other tribal groups; (3) the ethno-cultural *Volksgemeinschaft* called the "Armenians," both by the Persians and the Greeks, which was under the direct administrative control of the satrapal apparatus and is the most likely candidate to be associated with the *hay-k'* of later sources. Thus, one must conclude that during the Achaemenid Period the satrapy of *Arminal* Armenia was substantially larger than the ethnic territory occupied by "Armenians," and, therefore, the term "Armenian" would have had two different connotations at that time, one ethno-cultural and the other politico-territorial. That is, "an Armenian" of the Achaemenid period probably could be considered as *hay* in particular cases, but not necessarily always. At the same time, one should ask a legitimate question: to what extent did the ethnographically specific iconization of the exogenously defined titular "Armenians" in Achaemenid visual art contributed to the shaping of the unified endogenous *hayots* ethno-cultural identity?

Four centuries, sixteen to eighteen generations later, we are provided by one of the most significant testimonies concerning the Armenian ethnic history, which is given to us by Strabo, the ultimate Greco-Roman authority on Armenia of the Augustan epoch, who writes, concerning the ethno-cultural consequences of the political consolidation of the Kingdom of Greater Armenia under Artaxias I (*Artashes* – the phonetically Armenianized form of Old Persian *Artaxšaçā*/Artaxerxes), the following: ὥστε πάντας ὁμογλώττους εἶναι,[39] "...so that they all are of the same language." In the same section Strabo tells us that the Ararat Plain ("the country around Artaxata") was the core of Artaxias' expanding realm and, in the current absence of any reasonable alternative, one must conclude that it was the Armenian tongue, whose patchy presence in different parts of the Armenian Highland

---

[38] Xenophon. *Xenophon in Seven Volumes*, vol. 3. *Anabasis*. C. L. Brownson (transl.). Loeb Classical Library 90 (Cambridge: Harvard University Press; London: William Heinemann, Ltd., 1922).

[39] Strabo. *Geographica*, ed. A. Meineke, book 11, chapter 14, section 5. (Leipzig: Teubner, 1877).

since the beginning of the first millennium BCE, was attested in later Urartian cuneiform inscriptions,[40] that became the *lingua franca* within the borders of Greater Armenia. The conclusion that the Armenian tongue already had been the dominant language of Artaxias' heartland no later than the fourth century BCE, that is during the satrapal rule of the Orontid/Erwandid Dynasty, is unequivocally supported by the presence of the Old Persian loanwords in Armenian, from the times preceding the Macedonian conquest.[41] Together with the spread of the language, the endogenous Armenian, that is *hayots*, identity was assimilating other, smaller-scale ethno-cultural identities, yet never completing that process. One may suggest that during the Artashesid and Armenian Arsacid dynasties (189 BCE-428 CE) the endogenous *hayots* identity and the exogenous *Armenian* characterization became more synonymous than before, and also more than during several subsequent periods preceding the nationalistic Modern Age.

Powerful influences exerted by sociopolitical changes on perceptions of social identities are demonstrable through a salient example of the major divergence between the meanings of Armenia/Armenian and *Hayk'/Hayastan/ hayots* during the sixth through to the ninth centuries CE. *Geography* (*Ashkharhats'oyts*), a seventh-century CE text in Armenian attributed to the polymath Anania Shirakatsi, clearly presents the endogenous Armenian (*hayots*) perception of the gradual shrinking of the former realm of the Armenian Arsacid Dynasty in the course of the fifth through the seventh centuries, leaving at first sight a confusing impression that is also created by the existence of two recensions of the work (long and short) and inconsistencies within the multitude of extant manuscripts. Dividing the world into three "parts" (*bazhink'*), Europe, Africa, and Asia, each of which are subdivided into *ashkhark'* ("realms"), *Ashkharhats'oyts* includes 44 "realms" into Asia, beyond which lies the "unknown land."[42] "Realm 27" is Great Armenia (*Mec Hayk'*), composed of fifteen or sixteen *p'ok'r ashkhark'* ("little realms," usually translated as "provinces"), each of which, in their turn, are subdivided into a different number of *gawaŕk'*, "regions." Discrepancies between some manuscripts in the numbering of geographic units are immaterial for the purposes of this paper. What matters is to compare Greater Armenia to the descriptions of "*Veŕia that is Virk'*" (Iberia, modern Eastern Georgia), "*Aghbania that is Aluank'*" (Caucasian Albania, modern northern and western parts of the Republic of Azerbaijan), and Iran. The

---

[40] For a number of well-argued examples, see Gevork B. Djahukian, "Armenian Words and Proper Names in Urartian Inscriptions," in *Proceedings of the Fourth International Conference on Armenian Linguistics, Cleveland, Ohio, 1991*, ed. J. Greppin (Delmar, New York: Caravan Books, 1992), 49-59; Armen Petrosyan, "Armenian Elements in Urartian Language and Onomastics," Festschrift in Honor of Nicolay Harutyunyan, *Aramazd: Armenian Journal of Near Eastern Studies* 5, no. 1 (2010): 133-40.

[41] Antoine Meillet, "Sur les mots iraniens empruntés par l'arménien," in *Etudes de linguistique et de philologie arméniennes*, vol. 2 (Louvain: Imprimerie Orientaliste, 1911-12/1977), 149-50; R. Schmitt and H.W. Bailey, "Armenia and Iran, IV: Iranian Influences in Armenian Language," in *Encyclopaedia Iranica*, vol. 2, ed. Ehsan Yarshater (New York: Columbia University Press, 1986), 445-65.

[42] Ashot G. Abrahamyan, *Anania Shirakatsu matenagrut'yun: usumnasirut'yunə* [The Codicology of Anania Shirakatsi: A Study] (Yerevan: Matenadaran Press, 1944): 336-54.

territory of the latter has a more complex hierarchical subdivision than other "realms." The highest level in the geographical taxonomy of Iran is occupied by the four *k'ustk'*, each *k'ust* consisting of a number of *p'ok'r ashkhark'*, "little realms." The descriptions of both *Mec Hayk'* (Great Armenia) and Iberia include, among others, several *gawařk'* (regions): *Kałarjk', Artahank'*, and others, which, according to *Ashkharhats'oyts*, *"zor Virk'n unin"*, "are in the possession of Iberians." The same expression is used with regard to the regions that have become "the possessions of Albanians." The description of the northwestern *k'ust* of Iran (*K'ust-i K'apkokh*), which is simultaneously classified as one of the "realms" of Asia and one of the four highest subdivisions of Iran, includes at the same time the "realms" of "*Armn* (Armenia) – that is *Hayk'*, *Varčan* – that is *Virk'* (Iberia/Eastern Georgia), *Řan* – that is *Ałuank'* (Caucasian Albania), *Balasakan, Sisakan,* etc." *Sisakan*, which was, more or less, equivalent to *Siwnik'* (nowadays Syunik' Province of the Republic of Armenia), is defined as a "little realm" within the borders of Greater Armenia, but, at the same time, as a geopolitical entity separate from it and equivalent in rank to Greater Armenia within the political-administrative borders of the Iranian *K'ust-i K'apkokh*. R. Hewsen demonstrated, beyond reasonable doubt, that *Ashkharhats'oyts* presents a geographic image of *Hayk'* assembled from different historical periods,[43] most likely, one may think, emanating from the timespan that lasted from the Artaxiad Dynasty (second-first centuries BCE) to the repartitioning of the Roman Armenia by Justinian I into four parts in 536 CE, an image that was inserted into the geopolitical realities of the second half of the sixth (when *Siwnik'* was no more a province or "little realm" of Greater Armenia, but a separate "realm" of the *K'ust-i K'apkokh* in the Sasanian Iranian Empire)[44] and early seventh centuries CE.

This gradual diminution of endogenously perceived *Hayk'*, happening in the eyes of authors writing in Armenian language toward the time of the invasion by the Islamic Caliphate (around the middle of the seventh century CE), stands in an acute contrast with the exogenous perception of Greater Armenia during the same period. In 701 CE, the Umayyad Caliphate created an enormous province out of the territories conquered to the north of Mesopotamia, calling it *Arminia* and establishing its capital at Dwin in the Ararat Plain. Spanning over a territory by far exceeding that of the Armenian Highland, this new province included the countries named in Armenian language texts *Hayk', Virk', Ałuank', Chol* (*Darband* – Northeastern Caucasus, nowadays Dagestan of the Russian Federation), and the "lands by the Caspian Sea."[45] One may speculate that the creation of such a vast

---

[43] Robert H. Hewsen, *The Geography of Ananias of Širak: Ašxarhac'oyc', the Long and the Short Recensions*. Beihefte zum Tübinger Atlas des Vorderen Orients. Series B, Geisteswissenschaften, no. 77 (Wiesbaden: Dr. Ludwig Reichert Verlag, 1992).

[44] Tim Greenwood, "Sasanian Reflections in Armenian Sources," *e-Sasanika* 3 (2008): 8-9, http://www.sasanika.org/wp-content/uploads/e-sasanika3-Greenwood.pdf

[45] Aram N. Ter-Ghevondian, "Hayastani qaghaqakan inqnuruynut'yun VII dari 30-90-akan t't'" [Political Independence of Armenia during 630-690's], "Hay zhoghovrdi payk'ar Khalifayut'yan lci dem" [The Struggle of Armenian People against the Yoke of the Caliphate], in *History of the Armenian People*, vol. 2 (Yerevan: Publishing House of the Academy of Sciences of Armenia, 1984): 326.

territorial-administrative unit of the Caliphate had as precursor the short-lived unification and expansion of *Mec Hayk'* under the rule of Theodoros Rshtuni,[46] who took advantage of the geopolitical vacuum of power created by the collapse of the Sasanian Empire. When, in 652 CE, Theodoros Rshtuni established the first political contact with the Caliph's powerful Governor of Syria and Northern Mesopotamia, Muawiyah ibn Abu-Sufyan (the future Caliph Muawiyah I of the Umayyad Dynasty), acknowledging the suzerainty of the Caliph, he plausibly represented the ethno-political entity of *Hayk'* under the Middle Persian/Sasanian name *Armn*/Armenia, but the Umayyad province *Arminia*, patterned six decades later after the territory formerly under Rshtuni's control, had no ethno-political meaning. Moreover, as an Islamic administrative entity, that important province of the Caliphate was created, like others, on essentially anti-ethnic principles.

The aforesaid leads us to the conclusion that the historically attested, often occurring semantic difference between the endonyms *hay, hayots, Hayk', Hayastan* and the exonyms "an Armenian," "Armenian," "Armenia" must always be taken into account in studies of Armenian and/or *hayots* social identity and ethnic history. The question "what is meant by the word 'Armenian'?" must be inquired from a person or narrative using it as a qualifier in each particular case and is central to the research in Armenian identity.

Only keeping in mind the fundamentally important semantic distinction between the endogenous *hayots* identity and the exogenous characterization of "Armenian," we now may offer a few observations concerning the historical dynamics of the first, leaving the second to a further study on other occasions.

On the surface it may seem that a study of historical transformations of endogenous ethnic identities would overlap with a nationalistic view of the past, including, at least partially, an incorporation in the discussion of the intellectually discredited idea of so-called "ethnogenesis." But, if we take into account the aforesaid, we will immediately find the essential differences between the two. The core of the nationalistic view of the past hinges upon the ideas of a "primordial indigenousness" of a people in a given homeland, its superiority over other potential claimants over that territory through an assertion of time-depth, linear continuity of an ethnic culture, which politically leads to historical claims and the conceptualization of historical rights. One must acknowledge that individuals, social groups, peoples, nations, and even civilizations indubitably have rights of ownership with regard to their past, however, that has only an ancillary relation to social identity, a relation effectuated through such mechanisms as historical memory, real or imaginary, transmitted or reconstructed, which is an integral part of social identity. The concept of "ethnogenesis" also presupposes an unilinear development and structural stability of its outcome. Three typical approaches to the "study of ethnogenesis": two superficial and naïve, and one theoretically more elaborate have been developed in studies of Armenian past by scores of authors. Oversimplifying these theories for the sake of clarity, we can describe the first as a late nineteenth and early twentieth century migrationist interpretation of cherry-picked linguistic data and statements from ancient texts that lead to the erroneous conclusion

---

[46] Aram N. Ter-Ghevondian, "Hayastani k'aghak'akan," 312.

that the "Armenians" (a "formed people," one can assume) migrated from the Balkans and established themselves in the Armenian Highland, after the collapse of the state of Urartu. The second, which could be called autochthonist or primordialist, has been focusing almost exclusively on the "indigenous (presumably ethno-cultural) roots" of Armenians from the time immemorial. The third, elaborate theory of "ethnogenesis," assumed the existence of three separate antecedent components that form an ethnicity: the biological ancestry, an ancestral culture, and the language, which had conflated one with another under certain historical conditions, thus forming "the Armenian people." Even the third theory implicitly assumes that, at some point in the past, nowadays "Armenians" became "Armenians" and existed thenceforth as such throughout centuries and millennia, implying an ethno-cultural stability and adequate structural rigidity of the "Armenianness," which obviously is refuted by the aforementioned contemporary studies in social identity. These three theoretical approaches toward "ethnogenesis" never had explored the aforementioned question that is central for studies of identity: "what did 'Armenian' mean?" at different times and places, and in different socio-historical contexts.

Contrary to those theories, studies of social, including ethno-cultural and ethno-political identities, are not grounded in the concepts of continuity, linear transformation, and stability. As discussed above, they are derived from intertwined sets of distinctions and oppositions that constantly fluctuate and undergo changes, focusing on specific features, particularities, and diversity, rather than on generalities, uniformity, and continuity. And if, and when cases of continuity and persistence of identities are indeed observable, they are due to a persistence of distinctions, rather than linear developmental transformations.

This hypothesis is verifiable through the discovery of persistent sequence of sociocultural distinctions and oppositions that lead to the eventual shaping of endogenous Armenian, i.e. *hayots* identity, the sources of which, traced back for more than five thousand years, can neither be called Armenian, nor *hayots*, but the collective identities of the historical-cultural area reflected in the Kura-Araxes association of archaeological assemblages of the Early Bronze Age (ca. 3500/3400-2400/2300 BCE), which was spread over vast territories of what we call in Modern times the Armenian Highland, South Caucasus, and adjacent territories.[47] The majority of its components: from pottery repertoire and architectural designs of dwellings to multiple objects of visual arts representing aesthetic concepts, rituals, and mythology, express a startling contrast with functionally analogous cultural components known from the surrounding cultural-geographic areas. It ought to be duly noted that some major components of those assemblages, such as products of metalwork or small crude clay figurines do not partake in that specificity, being widespread beyond the geographic limits of the Kura-Araxes cultural traditions. However, that could not change its general assessment

---

[47] Karine Kh. Kushnareva, *Juzhnyj Kavkaz v IX-II tys. do.n.e.: Etapy kul'turnogo I sotsial'no-ekonomicheskogo razvitija* [South Caucasus in the 9th-2nd Millennia BCE: Stages of Cultural and Social-Economic Evolution] (Saint Petersburg: Peterburgskoe Vostokovedenie, 1993), 51-90; Mitchell S Rothman, "Explaining the Kura-Araxes," in *Fitful Histories and Unruly Publics: Rethinking Temporality and Community in Eurasian Archaeology*, eds. K.O. Weber, E. Hite, L. Khatchadourian, and A.T. Smith (Leiden: Brill, 2017), 236-44.

as a highly specific major historical-cultural areal phenomenon.

The Middle Bronze Age (ca. 2400/2300-1550 BCE) that followed witnessed a clear breakup with preceding agricultural traditions, caused, most likely, by a rapid aridization of the climate,[48] and continued fragmentation of the Kura-Araxes historical-cultural unity, which had begun already during the second half of the Early Bronze Age. Only very few cultural elements were traditionally preserved from the Kura-Araxes times, but instead a whole new system of archaeological assemblages emerged, which, nevertheless, again demonstrate a rapid development of cultural traits strikingly different from those that characterize the surrounding civilizations of Mesopotamia, Anatolia, Western Iran, and Northern Caucasus. Among those groups of artifacts that reveal a specific nature of cultural innovation during the Middle Bronze Age of the Armenian Highland, one should mention the *vishap* steles (vertical pillars erected on Alpine meadows and covered with reliefs depicting gigantic fish, bull hides with bucrania, storks,[49] regionally unique barrows (*kurgans*) with cultic pyres inside, a unique type of bronze weapons – long swords that, most likely, served as prototype for later Cretan and Mycenaean swords,[50] specific assemblages of painted pottery, which neither relate to the preceding Kura-Araxes ceramics, nor to the pottery production from neighboring areas of the Ancient Near East, etc.

The transition, ca. 1550-1450 BCE, from the Middle to the Late Bronze Age does not display a break in continuity as deep and comprehensive as the change that happened between the Early and Middle Bronze ages, which, most likely, was due to a more gradual process of sedentarization of nomadic pastoralists. Yet, the new conditions stimulated a consequent wave in cultural innovations that created a novel sociocultural pattern reflected in the content of archaeological assemblages. A new civilization of "cyclopean" fortresses, built of enormous rocks and surrounded by outer towns,[51] had risen during the Late Bronze and Early Iron Ages (ca. 1450-800 BCE), attesting to a spread of militarized bellicose societies that required a mass production of bronze weaponry. One would have to look at Mycenaean

---

[48] Gregory E. Areshian, "Les éleveurs non sédentaires et les civilisations agricoles à l'Age du Bronze sur le Plateau Arménien et au Caucase méridional," in *Nomades et sédentaires en Asie Centrale (Apports de l'archéologie et de l'ethnologie*, ed. H.- P. Francfort (Paris: Éditions du CNRS, 1990), 19-25.

[49] Alessandra Gilibert, Arsen Bobokhyan, and Pavol Hnila, "Dragon Stones in Context: The Discovery of High-Altitude Burial Grounds with Sculpted Stelae in the Armenian Mountains," *Mitteilungen der Deutschen Orient-Gesellschaft* 144 (2012): 93-132.

[50] Gregory E. Areshian, "Die Beziehungen der Kulturen des Armenischen Hochlands und der Zentralgebiete Südkaukasiens zu Vorderasien und der Ägäis in der Trialeti-Epoche," *Aramazd, Armenian Journal of Near Eastern Studies* 3, no. 2 (2008): 51-90.

[51] Raffaele Biscione, Simon Hmayakyan, and Neda Parmegiani, *The North-Eastern Frontier, Urartians and Non-Urartians in the Sevan Lake Basin: The Southern Shores* (Istituto di studi sulle civiltà dell'Egeo e del Vicino Oriente, Roma: CNR, 2002); Adam T. Smith, Ruben S. Badalyan, and Pavel Avetisyan, *The Archaeology and Geography of Ancient Transcaucasian Societies, vol. 1: The Foundations of Research and Regional Survey in the Tsaghkahovit Plain, Armenia* (Chicago: The Oriental Institute of the University of Chicago, 2009), 281-328.

Greece and, to some degree, to the Hittite Empire to find similarities in military architecture during the same period. Along with those, a new specific repertoire of bronze weapons, such as massive battle-axes with semi-circular edges of blades of the so-called "Amazonian" or "Transcaucasian" type, heavy cutting (as opposed to piercing) swords with rounded ends of blades, uniquely designed daggers, etc., were developed. Adding to the specificity of the historical-cultural area of the central South Caucasus and Armenian Highland of the Late Bronze-Early Iron ages are the assemblages of small bronze sculpture depicting ritual-mythological scenes and the highly standardized black burnished or dark gray pottery. Again, that rapidly developed cluster of indigenous cultural innovations has an appearance and, even to some degree, a functionality drastically different from the cultural production of the neighboring culture areas.

At the time, when the spotlight of locally written cuneiform texts (ninth-eight centuries BCE) illuminates the archaeological record of the Armenian Highland and South Caucasus, one would be astonished to see that the continuously recurring on the same territory specificity of the associations of archaeological assemblages, briefly sketched out in preceding paragraphs, overlaps with a large number of ethnic and sociopolitical names, some of which could be attributed with a substantial degree of certainty even to different language families (Indo-European, Hurri-Urartian, Semitic, Caucasian). That is not surprising, keeping in mind the upheavals, conflicts, and migrations of large and small ethno-social groups that created a new ethnic mosaic of the Near East and Mediterranean, following the collapse of the Bronze Age Empires in the twelfth century BCE. What may look surprising at first sight is that the archaeological record, also demonstrating visible changes, nevertheless does not reflect the degree of fragmentation that could have been assumed from the toponymics and onomastics of the region. It is seemingly more uniform than the linguistic pattern of the same period. Yet, the uniqueness of cultural characteristics of the Armenian Highland during the timespan between the end of the ninth and early sixth centuries BCE, was shaped through the integrative interactions between the creatively constructed imperial elitarian culture of the Urartian Empire[52] and the cultures of peoples conquered and absorbed by that empire. It was that process of imperial acculturation and restructuring that ended up shaping the sociocultural features, once again, differentiating the societies of the Armenian Highland from those of the adjacent macro-regions. It was the sociopolitical nature of the Achaemenid Persian Empire that allowed for the continuation of the mainstream Urartian societal traditions, after the subjugation of Armenia by Cyrus the Great in the middle of the sixth century BCE.

The spread of the Hellenistic civilization in Armenia, which began in earnest with the rise of Artaxias (Artashes) I to power after the battle of Magnesia in 190 BCE, was the divide marking the end of Ancient Near Eastern history in this geographic area. The apogee of Hellenization of Armenia could be placed within the first and second thirds of the first century BCE and attributed to the political and cultural program pursued by

---

[52] Paul E. Zimansky, "Urartian Material Culture as State Assemblage: An Anomaly in the Archaeology of Empire," *Bulletin of American Schools of Oriental Research* 299/300 (1995): 103-15.

Tigranes (Tigran) II the Great (95-55 BCE) and his son Artavazdes (Artavazd) II (55-34 BCE). That was the watershed after which Armenia transferred onto the pathway of the Mediterranean civilization a trend that accelerated with the periodic establishment of Roman domination, during the second and early third centuries CE. In this case, the breakup with the previous sociocultural traditions in Armenian history was manifested by the development of geopolitical and cultural reorientation of Armenia from the Iranian world toward the Mediterranean. Although the sociocultural system of Hellenistic Armenia featured many traits characteristic of other Hellenistic polities that existed within the sphere of influence of the Parthian Iran, it is exactly during that time that the symbolism reflecting the specific politico-cultural identity of the Artaxiad Empire and subsequent Kingdom of Armenia was developed. Our knowledge of that symbolism mostly is derived from Armenian and Roman numismatics[53] and from the monumental art of Orontid Commagene,[54] dynastically related to the Hellenistic Greater Armenia. Three essential interrelated components of that symbolic identity are readily identifiable: (a) the symbolic representation of the country of Armenia as a woman wearing a Phrygian cap, (b) the symbols of Armenian kingship consisting of the Armenian tiara derived from an Achaemenid prototype of crowns and modified to reveal the identity of particular rulers, together with a bow and a quiver filled with arrows, (c) a triadic group of supreme deities dominating the pantheon of the country, which included Oromasdes-Zeus, Mithras-Helios-Apollo, and Artagnes/Vahagn-Herakles-Ares. The origins and nature of other deities in the Artaxiad-Arsacid pantheon, including the two leading goddesses, Anahit and Astłik[55] known from a number of other sources, is not definitively established, as well as there are questions concerning their symbolism. Additional symbolism presented by Artaxiad (Artashesian) coins reveals the specific identities that an individual monarch desired to project. So, Tigran the Great identified himself as a Hellenistic city-builder, propagator of civilization and order harnessing the chaos of wild nature, which was conveyed by reproducing, on the reverse of his drachms and tetradrachms, the statue of Tyche, the Goddess of Good Fortune and patroness of cities, trampling under her feet the unruly Orontes River, sculpted for the newly founded city of Antioch by Eutychides at the

---

[53] Paul Z. Bedoukian, *Hayastani verabereal hṛomēakan dramner ew medalionner* [Roman Coins and Medallions Relating to Armenia] (Mkhit'arean Press, 1971). Paul Z. Bedoukian, *Coinage of the Artaxiads of Armenia*, Royal Numismatic Society Special Publication, no. 10 (London: Royal Numismatic Society,1978).

[54] Donald H. Sanders (ed.), *Nemrud Daği: The Hierothesion of Antiochus I of Commagene - Results of the American Excavations Directed by Theresa B. Goell*, vol. 1: Texts, vol. 2: Illustrations (Winona Lake, IN: Eisenbrauns, 1996).

[55] Scholars used to think that the worship of Anahit was introduced in Armenia together with the spread of Zoroastrianism, whether during the Achaemenid or Parthian periods, but currently there is a growing consensus that Anāhīd/Anaitis identified at a particular time with Greek Artemis spread from Anatolia and was worshiped in Western Iran before the adoption of Zoroastrianism as the official Achaemenid religion. See M. Boyce, M. L. Chaumont, and C. Bier, "ANĀHĪD" in *Encyclopaedia Iranica*, vol. 1, Fasc. 9, ed. Ehsan Yarshater (New York: Columbia University Press, 1988), 1003-11. http://www.iranicaonline.org/articles/anahid

end of the fourth century BCE. A different projection of Tigran's identity was designed for domestic consumption. His copper coins, which circulated in domestic markets, depicted Vahagn-Herakles as the symbol of invincible, brutal, and oppressive force. Artavazd II identified himself with Mithra-Helios, who was depicted driving his chariot on the reverse of the coins. This quite elaborate political and cultural symbolization of Armenia during the first century BCE-second century CE, combined with Strabo's aforementioned statement concerning the linguistic unity of Armenia, make a strong case in favor of the existence of an Armenian ethno-political identity during the Late Hellenistic and Roman periods. One may conclude, with a high degree of certainty, that at the time of collapse with the death of Artavazd II (34 BCE) of the Artaxiad imperial idea, the synonymy of "Armenia"/ "Armenian" and *Hayk'/hayots* was in the making. Obviously, Armenian identity of that time was not centered on Armenian ethnicity, rather it was shaped and nurtured by the sovereign authority of the country and focused on territoriality and the politico-cultural symbols stressed as identifiers by the ruling dynasty.

The next breakup with the preceding sociocultural traditions of the inhabitants of Armenia came with the adoption of Christianity as an essential part of the Arshakuni (Arsacid) nation-building project (see below), at the beginning of the fourth century CE. The disastrous consequences of compulsory Christianization to the Armenian pre-Christian identity and cultural heritage are well documented in original sources written in Armenian.[56] Yet, it is fascinating that, after this gap in developmental transformation of the Armenian culture, the latter was neither dissolved in the sociocultural environment of the larger Near East, nor assimilated by other Christian societies. The dynamics of deep history had revealed itself through the patterns of *longue durée*, once again creating a new system of interconnected cultural features that distinguished the "Armenianness" through differentiation and in opposition to the cultural production of adjacent areas. Salient examples of these newly developed cultural features are the Armenian alphabet created by St. Mesrop Mashtots, the gradual separation of the Armenian Church from and often in opposition to other Christian churches, the Armenian literary production of Late Antiquity (fifth-seventh centuries CE), the monumental prismatic stelae with relief depictions of Christian themes, the impressive gamut of different innovative architectural styles of church architecture,[57] and, later, the cross-stones – *khachk'ar*-s,[58] not to count other, less significant differentiating cultural features. All that was the result of major discretionary action, *Willkürbewegung*, and the following cultural and political collective drive, *Triebbewegung*, aimed at and resulting in the

---

[56] Agat'angełos, *Agat'angełay patmut'iwn hayots* [Armenian History by Agat'angełos], critical text prepared by G. Ter-Mkrtchyan and S. Kanayants, with translation into Modern Armenian and commentary by A. Ter-Ghevondyan (Yerevan: Yerevan University Press, 1983), 809-40.

[57] Jean-Michel Thierry, *L'Arménie au Moyen Age: Les hommes et les monuments* (Saint-Léger-Vauban, Yonne: Zodiaque, 2000).

[58] Anatolij L. Jakobson, *Armjanskie Khachkary* [Armenian Khachk'ars] (Yerevan: Hayastan Press, 1986); Hamlet L. Petrosyan, *Xačkar: Çagowmë, Gorçaŕowyfë, Patkeragrowtyownë, Imastabanowfyownë* [Khachkars: Origins, Functions, Iconography, and Semantics] (Yerevan: Printinfo, 2004).

creation of a new Armenian identity. Thus, the destruction of earlier identities followed by their replacement with new ones was becoming possible by means of continuous creation of new distinctions through innovative cultural production.

Thinking of that briefly described process, it is difficult to escape the conclusion that the historical dynamics of the Armenian identity, at least within its endogenous perspective, cannot be viewed as a linear developmental trajectory with its stages and periods identified on the basis of criteria that could be indiscriminately applied to every of those stages. To the contrary, one finds in that process a different kind of continuity through disintegration and fundamental transformation which could be described closely to the process of metamorphosis. Using this concept as a heuristic tool combined with the possibility of interpretation of biological and social processes as isomorphous, one may think of the aforementioned long sequence of sociocultural systems of the Armenian Highland, beginning from ca. 3500 BCE or earlier and continuing up to the middle of the first millennium BCE, as a sociocultural transformation without a predictable outcome.

The first metamorphosis developing from the transformation of the Urartian civilization, sometime soon after the sixth and before the first century BCE, resulted in the creation of the "primary ethno-cultural community" (i.e., the *Volksgemeinschaft*), with the autonyms *hayk', Hayk'*. Sociopolitical processes of consecutive imperializations during the Biainian/Urartian, Achaemenid, and Artashesian periods, together with the interplay of other sociopolitical and cultural factors had given that primary ethno-cultural community the status of a titular people, thus transforming it into an ethno-political entity.

The second metamorphosis of *hayots* identity was, beyond reasonable doubt, associated with and determined by the Armenian Arshakuni's nation-building project that originated as a sociopolitical response to the Sasanian revolution of 224 CE but remained incomplete because of the fall of the dynasty with the dethronement of Artaxias (Artashes) IV in 428 CE).[59] This transformed Arshakuni *hayots* identity was multicomponent and hierarchical: political-secular and religious-cultural at the same time.

The third metamorphosis was defined by the establishment of the Armenian Apostolic Church in the central position within the worldview of the *hayots* identity, which was derived from the acquisition of the political leadership by the Church in that ethno-political and cultural community between 428 and ca. 750 CE.[60]

The fourth metamorphosis was defined by two powerful sociopolitical factors that simultaneously had an impact on the *hayots Volksgemeinschaft*, resulting in the fragmentation of the latter. The first was the process of diasporization, which began in earnest from the eleventh century onwards. The second was characterized by the gradual disappearance of *hayots* polities of different ranks on the territory of the Armenian Highland. A transformation

---

[59] Areshian, "Sasanian Imperialism and the Shaping of Armenian Identity: Interdisciplinary Verification and Ambivalence of Empire-Nation Relationship," 146-63; C. Toumanoff, "ARSACIDS vii. The Arsacid dynasty of Armenia," *Encyclopaedia Iranica*, vol 2 (1986): 543-46, http://www.iranicaonline.org/articles/arsacids-vii

[60] Garsoïan and Mahé, *Des Parthes au Califat*; Garsoïan, *Interregnum*.

of different *hayots* ethno-political entities into ethno-cultural communities, still dominated by the Armenian Church and included on different terms in polities governed by "others" both in the Armenian Highland and in the diaspora, became the effect of these two factors. The profound effect exerted by the fourth metamorphosis on the collective identity could be best exemplified by the semantic shift in the meaning of its central concept. Up until the twelfth to the fourteenth centuries, the nominative case of the endonyms *Hayk'* (singular for the name of the country) and *hayk'* (plural for the name of people) had meant both the country and its people – *Volksgemeinschaft* and its genitive case *hayots* (assuming for a moment that it could be translated with the exonyms "Armenia," and "Armenians") would signify "Armenian," "of Armenia," and "of the Armenians."[61] Since that time, the word *hayots* is more and more often used with the meaning "of the Armenians," which is characteristic of the condition of diasporic ethno-cultural communities.

Finally, the origins of the **fifth metamorphosis** of the *hayots* identity could clearly be seen in the efforts of the sociopolitical and intellectual elites of the eighteenth century to re-create the nation-state.[62] A characteristic feature of this last phase in the sequence of metamorphoses of ethno-social identity is the strengthening of synonymy between the endonyms *hay, hayots* and exonyms 'an Armenian,' 'Armenian,' based on a more or less common ethno-cultural understanding, which today is conceived both by "insiders," "us" (i.e. Armenians) and "outsiders," "them" (i.e. non-Armenians) as something given. Among other salient features that emerged from this last metamorphosis of *hayots*-Armenian identity one could mention at least the following three. First is the destabilization of the central position of the Armenian Apostolic Church within the worldview of the *hayots*-Armenian identity, which began with the rise of Armenian secular political parties at the end of the 19th century. Second is the dramatic effect of the Armenian Genocide, which has been shaping the fifth

---

[61] This is only an abstract assumption used to simplify the understanding of this paper, since we know that, in the eleventh and fifteenth centuries, the endonym *Hayk'/Hayastan'* and the exonym "Armenia" were not synonymous. See the different meanings of "Armenia" conveyed in Persian poetry (Neẓāmi Ganjavi, ʿAlī-Šēr Navāʾī), or in the claims of the Saljuqid "Kings of Armenia" – Šāh-e Arman, or in Byzantine sources.

[62] For Israel Ori from Syunik,' see Valter Diloyan, "Hay azatagrakan payk'arə XVII dari erkrord kesin-XVIII dari aṙajin eresnamyakum [The Armenian Struggle for Liberation in the Second Half of the 17th-First Three Decades of the 18th Centuries], in *Armenian History*, vol. 3, eds. A. Melk'onyan et al. (Yerevan: Institute of History of the National Academy of Sciences of Armenia/Zangak-97 Press, 2010), 17-27; for Shahamir Shahamirean of Madras, India, see Khachig Tololyan, "Textual Nation: Poetry and Nationalism in Armenian Political Culture," in *Intellectuals and the Articulation of the Nation*, eds. R. G. Suny and M. D. Kennedy (Ann Arbor: University of Michigan Press, 1999), 79-105; Sebouh Aslanian, *Dispersion History and the Polycentric Nation: The Role of Simeon Yerevants'i's Girk' or Kochi Partavchar in the Armenian National Revival* (Venice: St. Lazarus, 2004); Vladimir B. Barkhudaryan, "Hayastanə ev hay azatagrakan sharzhumə XVIII dari veṙjin eresnamyakum" [Armenia and the Armenian Movement for Liberation in the Last Three Decades of the 18th Century], in *Armenian History*, vol. 3, book 1, eds. A. Melk'onyan et al. (Yerevan: Institute of History of the National Academy of Sciences of Armenia/Zangak-97 Press, 2010), 170-77.

metamorphosis to a very substantial degree. And the third is the unresolved fundamental conflict between the proponents of a modern, ethnically based Armenian state (the idea supported from Ori to the Armenian Revolutionary Federation-Dashnaktsutyun) and the developers of the concept of the Armenian nation-state, not based on a dominating ethnicity (the pathway pursued from Shahamirean to the authors of the current constitution of the Republic of Armenia). This conflict has not been yet adequately formulated, researched, and debated in the academic world and translated into political action, especially from the standpoint of understanding to what degree these two positions could be compatible.

At the end one may conclude that the diasporic Armenian identity of today is, by and large, ethno-cultural, still shaped by the worldview of the fourth metamorphosis. On the other hand, the Armenian identity in the Republic of Armenia has been transforming along the ethno-political pathway, characteristic of the fifth metamorphosis. In that we clearly see the opposition between the modern Armenian *Volksgemeinschaft* and *Gesellshaft*, the two mutually oppositional forms of societal systems investigated by Tönnies and Weber. Whether this opposition would result in a centrifugal differentiation of the current Armenian identities or their centripetal consolidation – only further research could probably find the answer.

THE FALL OF URARTU AND THE RISE OF ARMENIA[1]
Touraj Daryaee

Between 522 and 521 BCE, the Achaemenid Persian Empire faced a large number of uprisings in many of its satrapies, from North Africa to Central Asia. Darius, the Persian king, with the aid of his generals and those who were willing to work with him, was able to subdue these provincial uprisings.[2] Then, he set out to establish major reforms which resulted in the creation of a stable and functioning empire for the next two centuries. No doubt, in the midst of these uprisings, there were major alliances, personages, and groups that allowed Darius to become victorious at the cost of the enemies. Those who sided with the King of Kings were able to assert their own power in their satrapy (OP *dahyu-*), where they lived in and dislodged their opponents from power. In this essay, I intend to suggest that the supremacy of the Armenian people in the Caucasus was a result of their alliance with Darius. This resulted in the eclipse of the Urartian power in the satrapy which came to be known as Armenia, or in Old Persian *Armina*. Hence, *Armina* became a major satrapy and ally of the Achaemenids, which, in turn, provided stability for the Armenian people in the region of the Caucasus.

The tale of Darius' campaigns, lineage, and world-view is to be found in the trilingual inscription of Behistun, in Kermānšāh, written in Old Persian, Babylonian, and Elamite. It is here that, for the first time, we come across Old Persian word *Armina-* from Old Iranian *Arm-ina "(Area) with abandoned settlement sites" (Indo-Iranian *árma-* > Vedic *árma-* /

---

[1] I would like to thank Roberto Dan, Shervin Farridnejad, L. Khatchadourian and Shahrokh Razmjou for their kindness is supplying photographs, philological matters, and the Babylonian version of the Behistun inscription.

[2] For the chronology of the events and the revolts, see P. Briant, *From Cyrus to Alexander: A History of the Persian Empire*, ed. Pierre Briant (Winona Lake: Eisenbrauns, 2002), 116-18.

Avestan *arma-*).³ The inscription provides historical information on to the presence of the Armenians in the Caucasus and the political developments of the Near East, in the sixth century BCE. Before this time, the region, which the Armenians came to inhabit, was known in the Mesopotamian sources as *Uraštu* (^(KUR)*ú-ra-áš-ṭu*), i.e., Babylonian Urartu.⁴ In fact, in the trilingual Behistun inscription of Darius, when the Old Persian version discusses the events in the Caucasus and Mesopotamia, the term *Armina* is used, while the Babylonian version uses ^(KUR)*ú-ra-áš-ṭu*.⁵ This fact, may give us some ideas as to why this divergence exists between the Old Persian and the Babylonian version of the Behistun inscription. The region had been under Urartian domination for more than three centuries, and the subsequent association of the region and its people with Armenia, from the time of Darius onward, is an important event which is now receiving more important attention.⁶

Darius came to power through a dynastic struggle. Two versions of this ascendancy exist, Herodotus' version and the Behistun inscription, and depending on which version you follow, Darius either came to power after a pretender had usurped the throne, or he simply staged a coup and removed the rightful king.⁷ Either way, this action by Darius resulted in an uprising throughout the Persian Empire, from Central Asia to North Africa. In the narrative of the Behistun inscription, the Urartian-Armenian world appears as a point of contention and struggle for power on the northwestern edge of the empire. To quell the revolt, Darius sends two of his generals (OP *bandaka-*). The first is named Dadariši, against a revolting army (DB II 29-30):

θāti Dārayavauš xšāyaθiya Dādršiš nāma Arminiya, manā bandaka, ava madam frāišayam Arminam, avaθāšai aθanham paraidi, kāra haya hamiciya manā nai gaubatai avam jadi

Proclaims Darius, the King: (There is) an Armenian, Dādarishi by name, my vassal – him I sent to Armenia. Thus I said to him: "Go forth, (there is) an army which (is) rebellious (and) does not call itself mine – defeat that!"⁸

Dadariši met the rebels in Armenia, in a location called Zuzahya (Babylonian Zūza),

---

³ See R. Schmitt, *Wörterbuch der alpersischen Königsinschriften* (Wiesbaden: Reichert Verlag, 2014), 137-38.

⁴ V. Bănăţeanu, "Les Arméniens des inscriptions de Behistūn," *Studia e Acta Orientalia* 1 (1957): 65-81; and R. Schmitt, "'Armenische' Namen in altpersischen Quellen," *Annual of Armenian Linguistics* 1 (1980): 7-17.

⁵ E.N. Von Voigtlander, *The Bisitun Inscription of Darius the Great. Babylonian Version* (London: Corpus Inscriptionum Iranicarum, 1978), 25.

⁶ L. Khatchadourian, *Imperial Matter: Ancient Persia and the Archaeology of Empires* (Berkeley and Los Angeles: University of California Press, 2016).

⁷ For the Iranian narrative and the divergent tradition, see M.R. Shayegan, "Bardiya and Gaumāta: An Anchaemenid Enigma Reconsidered," *The Bulletin of Asia Institute* 20 (2006), 65-76.

⁸ R. Schmitt, *The Behistun Inscriptions of Darius the Great* (London: Old Persian Text, Corpus Inscriptionum Iranicarum, 1991), 57.

and by the favor of Ahura Mazda, Dadariši was victorious. Then again, a second battle took place at Tigra (Babylonian Digra), and again Ahuramazda bore Dadariši aid and the rebellious army was defeated.⁹ The third time the battle took place in Armenia in a location called Uyamā (Babylonian Ujama), and again by the aid of Ahuramazda, Dadariši was made victorious. Then a Persian vassal general (OP *bandaka-*), named Vaumisa, was sent to battle in the Caucasus. The first battle appears to have taken place at Izala (Babylonian Izalla), in Assyria, and then the second battle in Armenia, in a location called Autiyāra (Babylonian Utijari).¹⁰ By then, the Armenian and Persian generals awaited for the King of Kings himself to appear in Media before taking further action. Armenia appears to have been secured, while Darius moved into Media to take control of the heartland of the empire.

So why does the Babylonian version of the Behistun inscription mention Armenia as Uraštu / Urartu and the Old Persian as Armina / Armenia? I would like to suggest that what we are being privy to in the Behistun inscription is the turning point in the power-politics of the region and, more importantly, in the Caucasus itself. That is, the Old Persian version presents the immediate realities of the Caucasus between 521-518 BCE, where *Armina* and the Armenians were, or were on the verge of becoming , the masters of the Caucasus, while the Babylonian Uraštu offers a picture of the old power and traditional designation of the region, in the cuneiform tradition, for the past several centuries. So, we may suggest that the Old Persian version (*Armina*) projected the reality, while our Babylonian scribe, who knew the old designation and was not privy to the new reality of the late sixth century BCE, continued as usual to supply Uraštu.

Let us look at the names of the actors related to Armenia in the Behistun inscription. The rebel from Urartu-Armenia is named Arkha, son of Haldita. While the etymology of Arkha is unclear (where Ar(a)kh / Araxa, the suffix –khi/kha is most likely Hurrian-Urartian), Haldita, "Haldi is great," is theophoric with Haldi, the chief god of the Urartian pantheon.¹¹ Hence, Arkha, son of Haldita, has a strong association with the Urartian world. Not only the name exsits, but also toponyms with "Haldi" appears in the Urartian world, such as Haldei pātre, Haldiriu, Halitu.¹² One may suppose that it is indeed the Urartians who are against being brought into the newly upstart Persian Empire. The general of Darius, on the other hand, is named Dadariši. The etymology of the name stems from *darš-* "bold" or

---

⁹ Ibid., 58.

¹⁰ For the geographical identification of the places mentioned, see D. Pott, "Darius and the Armenians," in *Iranistik: Deutschsprachige Zeitschrift für iranistische Studien. Festschrift für Erich Kettenhofen*, vol. 5, nos. 1 and 2 (Teheran: Iran-Universitätverlag, 2006-2007), 134.

¹¹ M.A. Dandamaev, "Araxa," in *Encyclopaedia Iranica*, ed. E. Yarshater (Chicago: Chicago University Press: 1986), http://www.iranicaonline.org/articles/araxa-elamite-ha-rak-qa-akkadian-a-ra-hu-old-persian-form-of-the-name-of-a-leader-of-a-babylonian-rebellion-against-darius-i-db-3;

Dandamaev, *A Political History of the Achaemenid Empire* (Leiden: Brill, 1989), 123.

¹² I. M. Diakonoff and S. M. Kashkai, *Geographical Names According to Urartian Texts* (Wiesbaden: Dr. Ludwig Reichert Verlag, 1981), 38-9.

"daring" in Iranian,[13] but Darius designates him ethnically as an Armenian. Rüdiger Schmitt has observed that it is here, for the first time, that an Armenian person bears an Iranian name,[14] which means the name has also been Armeniani-ized, hence the close ties between the Armenians and Iranians.

Thus, the name of the rebel and the region described in the Babylonian version of the Behistun inscription mentions is Urartian, while the general and the region described in the Old Persian version of the Behistun is Armenian or has affinity with the Armenian world. Then, one can make the following assumption that the Armenians were immigrants to the region, if we are to follow Herodotus who calls them Phrygian colonists, whose names and dress was like the Medes.[15] They arrived at the time when the Urartians were at the apex of their power and settled in the region, perhaps as subordinates. Diakanoff has succinctly discussed the matter in terms of Phrygian and Armenian relations and believes that the Armenians, indeed, were a separate ethnic unit before the sixth century BCE.[16]

On the other hand, Urartu began to decline in power, but neither the Scythians nor the Medes in the seventh century BCE put an end to its power.[17] In fact, documentary evidence demonstrates that Urartu survived as a political entity, only to rise up against Darius.[18] One can suggest that the Armenians were able to achieve supremacy in the Caucasus in collaboration with Darius, and the Urartians who had staged a revolt against Darius and the Persians were defeated? This means the switching of position in Caucasus between the Armenians and the Urartians. As far as I know, there has not been any suggestions as to why the Urartians began to fade away, while the Armenians became the dominant group. My suggestion may be one possibility to explain the decline of Urartu and the rise of Armenia. One can also note that, based on the Behistun text, the Armenian ascendancy in the Caucasus can be dated between 20 May, 521 and 20 of June, 521 BCE, the dates of the three battles fought by Dadariši, the Armenian vassal general on behalf of Darius in Armenia.[19]

---

[13] R. G. Kent, *Old Persian Grammar, Texts, Lexicon* (Connecticut: American Oriental Society, 1953), 189.

[14] Schmitt, 57, ft. 29

[15] Herodotus VII.77; J.R. Russell, "The Formation of the Armenian Nation," in *The Armenian People from Ancient to Modern Times*, ed. R.G. Hovannisian (New York: Palgrave Macmillan, 2004), 24.

[16] I. M. Diakanoff, "Hittites, Phrygians and Armenians," in *Peredneaziatskiĭ sbornik, voprosy Khettologii i Khurritologii* (Moskva, 1961), 594-97 for the English summary. See also, G. Azarpay, *Urartian Art and Artifacts: A Chronological Studies* (Berkeley and Los Angeles: University of California Press, 1968), 117, ff. 242.

[17] R. Rollinger, "The Median 'Empire,' the End of Urartu and Cyrus the Great's Campaign in 547 BC (Nabonidus Chronicle II 16)," *Ancient West and East* 7 (2008): 60.

[18] Rollinger, 61. Piotrovsky only mentions at the beginning of the sixth century BCE that Urartu ceased to exist. See B. B. Piotrovsky, *Urartu: An Archaeological Adventure* (New York: Cowles Book Company, 1969), 199.

[19] M. A. Dandamaev, *A Political History of the Achaemenid Empire* (Leiden: Brill, 1989), 121-22.

Another question to pose is why the Armenians were so important for Darius? This is because, within the heartland of the Achaemenid Persian Empire, the most serious rival claimant to the throne was not far, in Media. Fravartiš was the most serious challenger, claiming to be the descendant of Cyaxares, rising nearby the Armenian region. Thus, having allies in Armenia in the north and Darius pushing from the west and from the south could have, in a sense, sandwiched and, eventually, crushed the Median rebels in the middle.

The number of the dead in Urartu also suggests that the violence in Armenia was far less than in other regions, for example the neighboring Media. According to the Babylonian Aramaic version of the Behistun inscription, on the Urartian front some 12,000 were killed or taken prisoner, while in Media it is stated that the dead were innumerable, and 18,000 people were simply taken as prisoners. The "innumerable" must have been in tens of thousands or more, as we have numbers mentioned for the dead in Margiana as 55,243.[20] Then, the province of Media was the most important area of concern, the highest deaths came from it, and it appears that Armenians, by working with Darius, were not only able to crush the Medes but were also able to crush the Urartians in Armenia. Hence, the old powers, the Medes and the Urartians, gave way to the new powers in the region, the Persians and the Armenians. That is why, when Darius commissioned his Behistun inscription, the scribes of the Old Persian and Elamite placed *Armina-* and *Harminuya-* for the province which the Babylonian version, as of habit placed *Uraštu.* This was the first event that signaled a change in the political landscape of the Armenian highlands and its connection to the Iranian world.[21]

Lastly, one should answer as to why among all of the Old Persian inscriptions which dominate the province of Persis and Media, there is only one outside of the Achaemenid heartland. This is in reference to the Xerxes' trilingual inscription in Armenia. The location and content of the inscription is very important to understand its significance. The inscription is placed on the ruins of the Urartian capital of Tušpa, on a blank niche on which Darius had already intended to log his achievement (Fig. 1). The inscription reads (XV 3):

ϑāti Xšayaršā xšāyaϑiya: Dārayavauš xšāyaϑiya, haya manā pitā, hau vašnā Auramazdāha vasai taya naibam akunauš, utā ima stānam hau niyaštāya kantanai, yanai dipim nai nipištām akunauš, pasāa adam niyaštayam imam dipim

---

[20] Ibid., 122; Briant, 118; Potts, 135.

[21] One can mention Xenophon's *Cyropaedia* or the *Education of Cyrus*. In that text, the childhood of Cyrus is given a fictional detail, and perhaps there is some truth to it. His close friend in *Cyropaedia*, during childhood, is an Armenian called Tigran. I believe here we are in the presence of a metaphor about the closeness of the Armenians and Persians from the beginning of the Achaemenid Empire, when the Armenian cooperation with the Persians resulted in their mastery over Caucasus, hence Armenia, at the cost of the Urartians. See, Ch. Nadon, *Xenophon's Prince: Republic and Empire in Cyropaedia* (Berkeley and Los Angeles: University of California Press, 2001), 77-86.

Figure. 1. Xerxes inscription at Tušpa (Courtesy of Roberto Dan)

nipaištanai, mām Auramazdā pātu hadā bagaibiš utāmai xšaçam utā tayamai kartam²²

Xerxes the king says: King Darius, who was my father, by the will/grace of Ahura Mazda built much good, and this niche he gave command to be carved, where he did not write and inscription, then I gave command to write this inscription, may Ahura Mazda together with other gods protect me and my empire and what has been built by me.

Khatchadourian has aptly discussed the significance of placing such a unique inscription outside of the Persian heartland, in a place that served as foundations of authority and in doing so incorporated a highland institution into the Persian world-view. It further

---

²² F.H. Weissbach, *Die Keilinschriften am der Achäemeniden* (Leipzig: Hinrichs, 1911), 116-19; S. Sen, *Old Persian Inscriptions of the Achaemenian Emperors* (Calcutta, 1941), 158-59; Kent, *Old Persian Grammar, Texts, Lexicon*, 152-53; P. Lecoq, *Les inscriptions de la Perse achéménide* (Paris: Gallimard, 1997), 263-64; M. Brosius, *The Persian Empire from Cyrus II to Artaxerxes II* (London: London Association of Classical Teachers, 2000), 64; A. Kuhrt, *The Persian Empire: A Corpus of Sources from the Achaemenid Period* (London and New York: Routledge, 2007), 301; R. Schmitt, *Die altpersischen Inschriften der Achaimeniden* (Wiesbaden: Reichert Verlag, 2009), 180-82.

Figure. 2. The Fortress of Tušpa (Courtesy of Roberto Dan)

demonstrated Ahura Mazda's reach to the farthest extent of the empire and the aspirations of the Achaemenid Empire.[23] I believe the inscription in Armenia also demonstrated this close relation and cooperation for the fate of both people in the sixth century BCE. Not only Darius was able to put an end to Urartian power in the Caucasus, but this accomplishment was marked at Tušpa, (Fig. 2), which brought the Armenians to power. Of course, it was the Armenians who were able to help Darius in defeating serious rebellions in Media and the Caucasus and were rewarded with their mastery over the province that came to be known as Armina / Armenia.

This connection in turn resulted in large-scale settlements of Persians in Armenia, be it through immigration, the nobility, or as exiles to the distant parts of the empire. The Persian imprint on this contact in Armenia could be seen from the use of Aramaic in official correspondences in the province of Armenia, and the worship of the Zoroastrian deity Anāhita, with its temple in Acilisene.[24] The close cooperation lasted for many centuries and

---

[23] L. Khatchadourian, "An Archaeology of Hegemony: The Achaemenid Empire and the Remaking of the Fortress in the Armenian Highlands," in *Empires and Diversity: On the Crossroads of Archaeology, Anthropology, and History*, ed. G.E. Areshian (Cotsen: Institute of Archaeology Press), 136-37; and L. Khatchadourian, *Imperial Matter: Ancient Persia and the Archaeology of Empires* (Berkeley and Los Angeles: University of California Press, 2016), 151-52.

[24] Briant, *From Darrius to Alexander*, 742.

through different dynasties, with the Orontids who began their career as Satraps of the King of Kings, holding on to the same name to make the first Armenian dynasty.[25] The break between Armenia and Iran came about only after the fourth century CE, with the adoption of Christianity by the Armenian Arsacid kings. However, by the end of Late Antiquity, the Armenians still supported the Persian kings and their closeness continued in history.[26]

---

[25] C. Toumanoff, *Studies in Christian Caucasian History* (Washington, D. C.: Georgetown University Press, 1963), 279-80.

[26] When the general Wahrām Čōbīn staged a mutiny against Khusrow Parwēz, he asked the Armenians to side with him, but they chose the Persian king over him. See T. Daryaee, "Wahrām Čōbēn the Rebel General and the Militarization of the Sasanian Empire," in *Studies on the Iranian World I*, eds. A. Krasnowolska and R. Rusek-Kowalska (Krakow, 2015), 197-98.

OF GOD AND LETTERS: A SOCIOLINGUISTIC STUDY ON THE INVENTION OF THE ARMENIAN ALPHABET IN LATE ANTIQUITY

Ani Honarchian

Two decades after Constantine the Great (280? -337 CE) allowed Christianity to be practiced freely in the Roman Empire, he decided to initiate a huge project. The Emperor ordered Eusebius, the bishop of Caesarea, to prepare fifty copies of the sacred Scriptures. In *Life of Constantine,* Eusebius explains that the Emperor wanted the books to be "written on prepared parchment in a legible manner and in a convenient, portable form, by professional transcribers thoroughly practiced in their art."[1] Constantine the Great thought that fifty copies of scripture would be enough for spreading Christianity in the entire Roman Empire. Of course, these books were about to be sent to churches in the major cities and were not meant for the public but, in the fourth century, hearing the holy words was good enough for the laity.

A century after this grandiose project, John Chrysostom, the archbishop of Constantinople, gave a sermon reproaching Christians for their lack of interest in reading the Holy Scriptures. He complains that, amongst the Christians, there was no one who had memorized even one of the Psalms, let alone read the gospels. Their excuse was that reading them was a job for monks. The archbishop thought it was unacceptable for people to say, "I am not one of the monks, but I have both a wife and children, and the care of a household." He wondered why laity supposed that reading the divine Scriptures appertained to only monks, when ordinary people needed it more.[2]

---

[1] Eusebius, Book IV, Chapter 36

[2] Homily ii. *On the Gospel of St. Matthew,* trans. in Nicene and Post Nicene Fathers. 1st series, vol. 10. 13. For further discussion, see Adolf Harnack, *Bible Reading in the Early Church,* trans. J. R. Wilkinson (London New York: Williams and Norgate, 1912), 118-9.

It is the case that most Christians were converted not by reading texts but by listening to the "Word of God" and mainly through the missionary efforts. Even though reading the gospels was what the Church fathers would have expected an 'ideal' Christian to do, it was not a reality, at least, not in the fourth and fifth centuries. Nevertheless, the spread of Christianity and its adoption as the new religion can be studied alongside the history of the book, literacy, and the creation of writing systems for languages such as Armenian, Coptic, and Syriac.

Many scholarly works have been done on the history of the invention of the Armenian alphabet and its origin. These works rarely place the issues of literacy and education within the bigger context of Late Antiquity. Most of the research done by scholars of Armenian studies suggests that the invention of the alphabet was prompted by religious and/or nationalistic motives. These works focus on questions, such as "when" exactly the alphabet was created and the importance of the alphabet for the Armenian literary culture.[3] This paper, however, explains "how" this alphabet was presented, justified, and understood in the backdrop of the fifth-century Late Antique world and within the Christian discourse of the period. I suggest that, regardless of what the origin of the alphabet was, that is more similar to Greek or Syriac, or when it was invented, the Armenian alphabet was presented as a sacred, awe-inspiring, distinctive gift from God to Armenians.

We cannot tell how Armenian speakers received the alphabet designed by Maštoc', but we know how this alphabet was advocated. This paper is a close reading of the life of the inventor of the Armenian alphabet, Maštoc', written by his disciple Koriwn.[4] I hope to demonstrate how, in Koriwn's work, the social structure of literacy, and the issues of identity, alphabet, and dialect intertwined to create a "promotional" discourse for the newly invented alphabet.

Some studies on the emergence of Armenian literary culture tend to apologize for the heavy Christian discourse that colors the account of the author of the *Life of Maštoc'*. For instance, one scholar complains that when Koriwn tells us about the creation of the alphabet, his "history" is infused with a mixture of "devout exuberance" and "a sense of victory."[5] In the introduction of the English translation of *Life of Maštoc'*, the translator, Bedros Norehad, encourages the "average reader" to ignore the beginning paragraphs, which are filled with "biblical allusions and quotations."[6] Nersoyan complains that Koriwn is not

---

[3] Artašes Martirosyan, *Mashtots', Patmakan Tesut'yun* (Yerevan: Historical View, 1982); M. Minasian, "Koriwni grk'i 'ams erkus' ev verjabane" [The "Two Years" of Koriwn's Book and its Epilogue], *Handes Amsoreay* (1983): 7-12; Hagop J. Nersoyan, "The Why and When of the Armenian Alphabet," *Journal of the Society for Armenian Studies* 2 (1985): 51-72.

[4] Manuk Abełian's critical edition is generally preferred by scholars. My references are from Abełian's edition, and I used Bedros Norehad's translation.

[5] Nersoyan, 51-72.

[6] Koriwn, *Life of Maštoc'*, trans. by Bedros Norehad (New York: Armenian General Benevolent Union of America, 1964), 7.

detached from the event, as an objective historian.⁷ Being dipped in a religious mindset is not unique to the Armenian historical accounts: it can be traced across literary cultures of Late Antiquity. Indeed, these are very important elements that deepen our understanding of the how of the Armenian alphabet was presented.

Averil Cameron, who has studied the place of Christian language within the social and political rhetoric of the Roman Empire, argues that Christianity had a powerful and totalizing discourse, flexible enough to be understood by the public and sufficient enough as a political instrument. Cameron explains, "Christian discourse provided both the framework within which most people looked at the world and the words that they used to describe it."⁸ This powerful discourse should be studied in Koriwn's work, as he explains the significance of the alphabet for the Armenian-speaking people. Koriwn's depended on this discourse to convey his message. Early Christianity was not only a matter of ritual or ethical behavior, it was a matter of teaching, interpretations, and the definition of a discourse. Koriwn tried his hand at Christian literary discourse, and there is no need to apologize for his heavy use of religious elements. The current of Christian discourse world of Late Antiquity that Koriwn was aware of offered something new to men like him: an intellectual universe, where matters could be both understood and explained. Koriwn writes:

> For the land which had not known even the names of the regions where all those wonderful divine acts had been performed, soon learned all the things that were, not only those that has transpired in time but that of the eternity which had preceded, and those that had come later, the beginning the end and all divine tradition.⁹

Peter Brown explains how a Christian man found the concept of "God of the Universe" appealing, as a way to explain the whole universe. Men like Augustine, John Chrysostom, Basil of Caesarea, and Symeon the Stylites, who believed that Christianity could provide an intellectual universe, and that, by belief, the unknown world would make sense, marked the world of Late Antiquity.¹⁰ It was the belief held by men like Koriwn about men like Maštocʻ, who acted as the servants of this "One God" and were dependent on the supernatural for guidance and direction. The, "New mood" of Late Antiquity shaped the writing of Koriwn, who truly believed that, by the grace of the Armenian alphabet, a door would open to Armenians, a point of entry to a bigger, more sensible world. The moment Armenians understood this

---

⁷ Ibid., 51. About the writing style of Koriwn, see Edward Mathews, "The Life of Maštocʻ as an Encomium: A Reassessment," *Revue des Études Arméniennes* (1993), 5-26; Abraham Terian, "Koriwn's Life of Maštocʻ as an Encomium," *Journal of the Society for Armenian Studies* 3 (1987): 1-14.

⁸ Averil Cameron, *Christianity and the Rhetoric of Empire: The Development of Christian Discourse* (Berkeley and Los Angeles: University of California Press, 1991), 222.

⁹ Koriwn, Chapter XI.

¹⁰ For more on the emergence of the holy men in Late Antiquity, read Peter Brown, "The Rise and Function of the Holy Man in Late Antiquity," *Journal of Roman Studies* 61(1971): 80-101.

"Biblical world," the knowledge of the past and the future would be revealed to them.

How the Alphabet was Presented: Distinct and Sacred

The question I would like to answer here is why an alphabet had to be invented for a religion whose members were not expected to actually read its Holy Book? And why would it have the support of both the court and the Church of Armenia? On the Sasanian side of Armenia, Vramšapuh (reg. 401-417) supported the development of the Armenian alphabet, and, on the Roman side, the approval of Theodosius II (401-450) was obtained.[11] Even though the Sasanian King Yazdgerd I (r. 399-420) was not directly involved with the innovation[12] and is not even mentioned in the *Life of Maštoc'*, Yazdgerd's peaceful reign made the traveling of Maštoc' between the Greater Armenia under the suzerainty of the Sasanians and Lesser Armenian under the Romans possible.

The creation of the Armenian alphabet was not solely to represent the sounds of the Armenian language. With some difficulty and adjustments, Syriac or Greek scripts would have conveyed the sound system of the Armenian language.[13] Armenians probably were using other languages and their respective scripts in different settings. Middle Persian might have been used in communication with the Sasanian court, and Greek and Syriac were necessary to deal with religious matters. To understand why a whole new alphabet was invented for the Armenian language, we must turn to sociolinguistic studies. Traditionally, the field of sociolinguistics has focused on the influence of the social environment on spoken language. More recently, sociolinguistics has been turning its attention to written language as well. Peter Unseth, who has studied the motivations behind choosing a national writing system, believes that factors, such as ethnic and national identities, political movements, perceived prestige, and religion are factors that influence why a community of speakers choose to adopt a certain script.[14] Similarly, William Smalley, who was involved with creating a writing system for the Hmong language, thinks that cultural, social, political, and ideological factors each play a major role in the acceptance or rejection of a writing system, even more so than linguistic matters.[15] The use of a particular script for a specific language, for instance Arabic script for Persian or Latin for Turkish, are conventional choices and subject to change. Some languages use several scripts as a norm, and their users have a certain amount of flexibility in choosing between them. As Unseth points out, Serbs (Orthodox) use Cyrillic, and Croats

---

[11] Koriwn, Chapter XVI.

[12] Krikor H. Maksoudian, *The Origins of the Armenian Alphabet and Literature* (New York: St. Vartan Press, 2006), 58-9.

[13] See Sebastian Brock, "Armenian in Syriac Script," in *Armenian Studies. Études arméniennes. In Memoriam Haïg Berbérian*, ed. Dickran Kouymjian (Lisbon: Calouste Gulbenkian Foundation, 1986), 75–80. Also Hidemi Takahashi, "Armenian Garshuni: An Overview of The Known Material," *Journal of Syriac Studies* 17 (2014): 81-117.

[14] Peter Unseth, "Sociolinguistic Parallels Between Choosing Scripts and Languages," *Written Language & Literacy* 8, no. 1, (2005): 19-42.

[15] William A. Smalley, "How Shall I Write This Language?," *The Bible Translator* 10 (1959): 49-69.

(Catholic) use Roman script.[16] The choice does not depend on a linguistical factor but on a social value of the script. This brings us back to Maštocʻs motivations for shunning an already available script to create an original script. Instead of choosing a side with which to merge, i.e. Greek (Rome) or Syriac (Persia), he endeavored to maintain a distance for the speakers of the Armenian language by inventing a distinct, original script for Armenians.

Koriwn explains that the first attempt of Maštocʻ was to work with the alphabet suggested by King Vramšapuh. This alphabet was discovered by Daniel, a Syrian bishop. However, it was rejected by Maštoc, after he tried teaching the letters to some young students. Koriwn believes that the alphabet was insufficient to form Armenian syllables and adds that it was an alphabet "buried and then resurrected from other languages, (*aylocʻ dprowtʻeancʻ*).[17] Apart from the linguistic issue of representing the sounds of Armenian speech, the alphabet was not uniquely Armenian, it had been used to represent another language.[18] It was not "*distinctive*," a factor that sociolinguistic studies find essential for a positive reception of a new alphabet.[19]

According to Koriwn, Maštocʻ, in his quest for the perfect alphabet, went to Samosata, where a scribe of Greek letters, a Syrian man by the name Ropanos, helped him with the design of the alphabet. The visual similarity of the Armenian script with Greek could be the result of this scribe's efforts, not to mention that the order of the Armenian alphabet is following Greek. The desire to have a script graphically similar to one language or dissimilar to another, logical or not, are social factors that influence the success of a writing system. The Greek alphabet was used in communication with the Roman Church, and resemblance with it could have been considered prestigious. However, the goal was to have a *distinctive* Armenian alphabet. Koriwn emphasizes this aspect, which was in accordance with the biblical image of where Armenian people belonged in the world. Koriwn states,

> I had been thinking of the God-given alphabet of the Azkanazian nation and of the land of Armenia- (*Zazkanazian azgi yev zhayasdan ašxarin asdvasta-pargev*) when in what time, and through what kind of man that new divine gift had been bestowed, as well as luminous leaning and angelic.[20]

The first point Koriwn refers to as a binding element between all the Armenians is having a common ancestry (*Azchanazi*): a biblical understanding of the place of Armenia

---

[16] Unseth, 24.

[17] Koriwn, Chapter VI.

[18] See James R. Russell, "On the Origins and Invention of the Armenian Script," *Le Museon* 107 (1994): 317-33. Russell argues, without much evidence, that the language known as Danielian could have been modeled after the Manichaean alphabet.

[19] Nersoyan studies in length how Syriac alphabet was rejected as the proper writing system for the Armenian language. He argues that issues, such as the fear of the influence of Syriac speaking Church of Persia, were more important than the difficulty of matching the alphabet with the Armenian syllabic system. Nersoyan, 57-8.

[20] Koriwn, Chapter I.

(*Hayasdan ašxar*). Koriwn speaks about the Armenian alphabet as a divine gift given to the Azchanazian nation. To refer to the Armenian people as Azchanazian is based on a passage in the Bible that says, "Call together against her the kingdom of Ararat, Minni, and Ashchenaz" (Jeremiah 51:27).[21] Notice also how, Koriwn points to the divine nature of the Armenian alphabet connecting God to the land and the Armenian people.

Considering a script sacred, magical, holy, and God-given is not unique to Armenians. Ancient Egyptians called their script *mdw·w-nṯr* (*medu-netjer*) "god's words." This monumental art system was never used for any other language but Egyptian. In the rabbinic tradition, the Hebrew script was believed to go back to the creation. In the Mishnah, it is stated that the letters of Hebrew alphabet, as well as the art of writing, were created on the sixth day.[22] Letters were thus created right before the creation ended, at the end of the sixth day. According to the Jewish tradition, it was God himself, with his finger, who wrote on the stone tablets given to Moses on Mount Sinai. God is actually keeping a heavenly book, inscribed with people's names, which he adds or erases, thereby inscribing people with their eternal fate. In the Islamic tradition, two angels, *kirāman kātibīn* ("two noble writers"), are responsible for recording a person's action and thoughts: "Over you stand watchers, noble recorders who know what you do..."[23] The Qur'an, as a Holy Book, has a significant place in Islam. Michael Cook states that the Qur'an is not only a respected book of Muslim religion but it is a sacred object, a sacred codex. The unbelievers should not touch the Qur'an, and the majority of Muslim warriors would not take the Qur'an with them to a non-Muslim territory, dreading that it might fall into the hands of non-Muslims.[24] Such understanding of writing systems, as holy mysteries, is typical of largely oral societies.

According to Koriwn, the Armenian alphabet will be the tie between Armenian lands, the people, their language, and the blessing message of God. A unified Armenia across Roman and Sasanian borders, was the ideal scenario of the Armenian Christian historiography, however it was not a reality, after Armenia was divided between Rome and the Sasanian Empires in 387 CE, and surely not when Maštoc' started his mission. This fact does not invalidate the importance of the conceptual existence of boundaries for a unified Armenian territory. Robert Thomson argues that the development of Armenian literature solidified the ties between the Armenians on both Roman and Iranian sides. Furthermore, in Arsacid Armenia, there were some fifty noble families of varied size and power, each with their own military forces, and it seems that the voice of church authorities spoke for broader interests

---

[21] In the genealogies of the Hebrew Bible, Ashchenaz was the first son of Gomer (Genesis 10:3, 1 Chronicles 1:6), and Gomer was the grandson of Noah through Japheth.

[22] Mishnah, Pirkei Avot 5:6. Trans. Jacob Neusner, *Torah From Our Sages: A New American Translation and Explanation = Pirke Avot*. (Chappaqua, N.Y.: Rossel Books, 1984. It is not possible to date this section of Pirkei Avot, because it was not attributed to any rabbi but most scholarship argue for a third century dating.

[23] *Al-Infitar* (Torn Apart) 82: 10-13, English Translation: *The Qur'an*, trans. M.A.S. Abdel Haleem (Oxford University Press, 2005), 412.

[24] Michael Cook, *The Koran: A Very Short Introduction* (Oxford: Oxford University Press, 2000), 59-60.

within the Armenian territories.[25]

Standardization of a language

The Armenian language which Maštocʻ was committed to put into writing was one particular dialect spoken by Armenians. Most likely, it was the dialect spoken in the court of the Arsacids. Koriwn refers to some other dialects of Armenian in passing. What Maštocʻ was doing was standardizing one dialect of Armenian as the proper Armenian language, and that task was not easy. Koriwn, here, explains the attitude of Maštocʻ toward the people of the "region of Medes" who lived near the borders of the Sasanian Empire. Their dialect was described as being harsh in manner because of their crude and corrupt dialect. (*xecʻbekagoyn ew xošoragoyn lezowin džowaramatoycʻkʻ ēin*). Koriwn says it was through the instruction of Maštocʻ, who took it upon himself to refine them, that,[26]

> ... they made them, offspring of many generations, intelligible, eloquent, educated, and informed of godly wisdom. Thus, they became immersed in the laws and commandments to the extent of becoming distinguishable from their fellow natives.

Linguistic standardization requires political and social control over writing that plays as an administrative and communicative tool. In order for this tool to work, it has to be unified across different dialects of a language. After all, as Max Weinreich once said, "a language is a dialect that has an army and a navy." In the case of the Armenian language, we should add a cross to this equation. The case of the creation of Georgian alphabet, where many dialects and languages were spoken by the population, is a great example of how one writing system built and bounded together a "one people."

And thus they who had been gathered from among so many distinct and dissimilar tongues, he bound together (*mi azgi kabial*) with one divine commandment, transforming them into one nation and glorifiers of One God.[27]

Social Setting of the Written Word

In *Ancient Literacy,* William Harris argues that it was accepted that the lay poor could remain illiterate, and it was not the concern of the church to spread education beyond its established boundaries. Harris suggests that crucial factors, such as industrialization, urbanization, the rise of elementary schools, and the involvement of the state with the education of the public were all crucial elements that contributed to mass literacy. These factors were not in place before the eighteenth and the nineteenth centuries, and even a conceivably strong ideology such as Christianity, with its focus on books, was not robust

---

[25] Robert W. Thomson, "Armenia in the Fifth and Sixth Century," in *The Cambridge Ancient History. Empire and Successors, Ad 425-600 Vol. 14*, ed. Averil Cameron, Bryan Ward-Perkins, and Michael Whitby (Cambridge: Cambridge University Press, 2008), 667.

[26] Koriwn, Chapter X.

[27] Koriwn, Chapter XV.

enough to mitigate for the lack of these factors.[28]

Even one century after the conversion of the Armenian royal court, Christianity was functioning and spreading in the Armenian language without the need for an alphabet. However, by the fifth century, the Church had developed a widespread structure of ecclesiastical authority in which literacy was necessary. Matters relating to the discipline, the hierarchy, and the organization of the Church were amongst the concerns of the first Council of Nicaea, in 325. For this ecclesiastical structure to function, its members had to communicate. To this end, literacy was crucial. Literacy became a means for the Church to get involved and regulate the mores and daily affairs of the community: to send orders and to demand reports. If another medium was available, the Church would have used it to circulate its directives and news.[29]

One of the main contributions of Maštocʻ to the Church was to make orders for monks to learn how to read and write.[30] He established places for this and addressed the Church's need for a functional communication network in Gołtn an Swinik and other regions. Koriwn says that, in Siwnik, Maštocʻ gathered the youth from more "brutal, barbarian, and fiendish regions and cared for them and instructed them as a teacher, educated and advised them, ordained bishops (*episkopos tesowčʻ ekełecʻwoyn*) as overseers from among those barbarians and filled the region with the monastic order."[31] Maštocʻ left behind overseers who, besides their evangelical mission, could report news and execute orders. To maintain the integrity of the Church, ecclesiastical hierarchies depended on the exchange of letters between them. Koriwn explains that the mission was mostly to preach the gospels and establish orders for monks. These orders for monks became the institutions that spread Christianity and educated monks who would have remained illiterate otherwise. It is not very clear if this setting was structured to teach laymen as well. This supports Harris' theory that the Church's was not interested to teach literacy beyond some selects groups.

The hierarchy of groups that received the new alphabet – the king, a section of the army, and one prominent noble family – shows the importance of writing for certain groups of fifth-century Greater Armenia. The groups that received or should have received literacy, according to Koriwn, are the royal court and the entire *azatagound banak* (aristocratic entourage), the men of the Mamikonean family, the royal garrison, and later, the regions that were still pagan.[32]

While the invention of the Armenian alphabet made reading and writing in the Armenian

---

[28] William V. Harris, *Ancient Literacy* (Cambridge: Harvard University Press, 1989), 30.

[29] Robin L. Fox, "Literacy and Power in Early Christianity," in *Literacy and Power in the Ancient World*, ed. Alan K Bowman and Greg Woolf (Cambridge: Cambridge University Press, 1996), 148.

[30] See Nina Garsoïan, "Introduction to the Problem of Early Armenian Monasticism," *Revue des Études Arméniennes* 30 (2005/07): 177-236. Garsoïan has shown that, during the fourth and fifth centuries, there were no settled and permanent monastic communities. Ascetic groups were held together by a common covenant or purpose at the service of a particular saint or a shrine.

[31] Koriwn, Chapter XIV.

[32] Ibid., Chapter XII.

language possible, it did not result in an immediate spread of literacy. It is hard to achieve literacy, even with an alphabet designed for your native spoken language. Learning the shape of the letters and their respective pronunciations was quite possible, but becoming literate demanded years of education, and without social value attached to the practice of literacy, there was little motivation for its acquisition.

Conclusion

The most difficult task that was set in front of Koriwn was to to convince a group of language speakers to agreed that the writing system that was created by Maštocʿor bestowed upon him was the one true representation of their language. This, of course, was not possible without the support of both the Court and the Church. Similar to the situation in the Roman Empire, in Armenia, the spoken word must have been integral for the spread of Christianity among the population. Most Christians were converted not by reading texts but by listening to the "Words of God." Koriwn was adamant that Moses, Paul, and the gospels of Christ became *hayaxos* (Armenian-speaking), not just written down in Armenian. In this essay, I have argued that in order for an alphabet to be accepted, regardless of its linguistic capacity, its cultural aspects need to be emphasized. Koriwn presents the alphabet as a sacred gift, given by God to Maštocʿ, a thread that tied one God to one people and one land. The standardization of the Armenian language was asserted by Maštocʿ, who believed that the dialectal form of Armenian he spoke must be taught to the monks whose dialect was not deemed acceptable .

The style of Koriwn's writing, filled with biblical imagery and exaggerations, was common in Late Antiquity. It is true that Koriwn's writing was hyperbolic: he even placed the alphabet and Maštocʿ at a higher level than the tablet of the covenant and Moses. He held Maštocʿ in high regard, but his account was more focused on elevating the status of the alphabet for Armenians. By attributing properties such as sacredness and distinctiveness to the alphabet, which was more inclined to Greek in its style, Korwin tried to make it as attractive as possible.

THE REBELLION OF BABAK AND THE HISTORIOGRAPHY OF THE SOUTHERN CAUCASUS

Khodadad Rezakhani

Bringing together regional history and trans-regional/global history is among the most difficult tasks of modern historians. The first type of history is the forte of specialists of often obscure, particular languages or that of archaeologists, while the second type is practiced by those better equipped with methodologies of sociology, anthropology, or economics. Regional specialists have difficulty letting go of their sources and avoid broad brushstrokes over their region's characteristics, while global historians, seeing similarities and connections across cultures and polities, see it necessary to highlight these aspects. Of course, it is natural to assume that both these sensibilities can and should be considered within any historical study, but the fact is that methods of achieving this are still in their infancy and the gap seems, astonishingly, unfordable at times. This is despite successful attempts, particularly in the context of European history.[1]

In studying the history of the non-European world, in this case West Asian/Middle Eastern history, one commonly encounters this same issue, particularly when ancient and medieval history is being considered. The approaches by the "Orientalist" and "historian" camps often make it impossible for the two to communicate or exchange ideas on very similar subjects. It also results in ignoring certain episodes by each side or even treating the same

---

[1] Perhaps a good example of this is Chris Wickham, *Framing the Early Middle Ages: Europe and the Mediterranean, 400-800* (Oxford: Oxford University Press, 2005). In relations to Iran and the non-Western world, some of the more comparative approaches have been attempted by Josef Wiesehöfer, "The Achaemenid Empire," in *the Dynamics of Ancient Empires*, eds. Ian Morris and Walter Scheidel (Oxford: Oxford University Press, 2009), 66-98; Touraj Daryaee, *Sasanian Iran: The Portrait of a Late Antique Empire* (Costa Mesa: Mazda, 2008).

events differently. A case to this point is the history of the southern Caucasus as studied by specialists of the history of this region and based on local (often Armenian or Geogrian) sources,[2] as opposed to the way it is studied by those concerned with Iranian or Islamic history.[3] A particular episode of this history, that of the Rebellion of Babak Khurramdin (AD 816-837), is the concern of the present article. The goal here is to consider not only all sets of sources on this event and, in the process, shed light on regional and transregional importance and context of the rebellion, as well as pleading for its consideration as part of the wider political events of the Islamic World. This, specifically, is the context of regional interactions between the Dailam/Caspian Coast region and the area of *Arminiyah*, the Islamic region that included the entire southern Caucasus and upper Mesopotamia.[4]

Introduction

Among the many peculiarities of the historiography of the southern Caucasus is its strict division among ethnic lines, lines that go beyond even borders and have more to do with perception than reality. Thus, those working on "Armenian" history see the entire region as Armenian, and those working on "Iranian" or Islamic history see it as a northern "appendage" of the Iranian world. Historians of "Azerbaijan" or "Georgia" or "Turkey" follow suit in similar manners, with a particular space reserved for Pan-Turkic historiography.[5] Within this fragmented scheme, certain episodes of history fall into the fault lines, often studied within a single historiographical tradition and in disregard for others. A great part of this is owed to the problem of languages and sources. Islamic sources, written in Arabic or Persian, treat the events in this part of the world as matters of peripheral importance, often continuing a tradition of seeing the entire region as one of conflict between the powers in Iran and those centered further west, mainly Rome/Byzantium. Armenian and Georgian sources, on the other hand, speaking from their regional focus, give pride of place to local events, and in familiar terms to other historiographies, present local actors as major agents of the wider regional history.[6] This latter treatment is, at times, astonishingly short-sighted

---

[2] See Cyril Toumanoff, *Studies Christian Caucasian History* (Washington DC: Georgetown University Press, 1965); Robert Hewsen, *Ethno-History and the Armenian Influence upon the Caucasian Albanians. Classical Armenian Culture: Influence and Creativity* (Philadelphia: Scholars Press, 1982).

[3] To be fair, the great Vladimir Minorsky is a significant exception to this rule. See among others, V. Minorsky, "Caucasica IV," *Bulletin of the School of Oriental and African Studies* 15, no. 3 (1953): 504-29. Ahmad Kasravi too used Armenian sources, although in somewhat shallow ways, in his series on minor dynasties of northern Iran. See A. Kasravi, *Shahriyārān e Gomnām* (Tehran: Jami, 1977 [1307])

[4] Robert Hewsen, *Armenia: A Historical Atlas* (Chicago: University of Chicago Press, 2001).

[5] For a survey of many of these, see Harun Yilmaz, *National Identities in Soviet Historiography: The Rise of Nations Under Stalin* (London and New York: Routledge, 2015), 38-48.

[6] In a way, this is consistent with many regional histories that come to existence in medieval Iran, including *Tarikh-e Sistan*, where the entire history of pre-Islamic Iran is explained from the point of view of Sistan and its heroes, most significantly Rostam. See Mohammad-Taqi Bahar, ed., *Tarikh-e Sistan (the History of Sistan, Written between 445-725 AH)*. 2nd. ed. (Tehran: Moin, 1387).

and uninterested in wider contexts, such as those mentioned in the Islamic sources. While some historians have had the broader aim of relying on both sets of sources and drawing on them to present a more complete picture of regional history,[7] still many have stopped short of considering the entire picture and have left fascinating details of history unexplored.

Perhaps one of the most interesting of these is the history of the local relations between the south Caucasus and the regions to its south and east, namely Azerbaijan/Aturpatgan (as in the historic region, not the modern nation-state), Shervan (in the modern nation-state of Azerbaijan), and Dailaman (part of the modern Iranian province of Gilan). In terms of local Armenian and Georgian historiography, these regions and their political history is often treated as foreign (often Muslim backed) intrusions into local affairs, something to be despised and resented and dismissed. For Islamic historians, both medieval and modern, the inter-regional relations in this part of the Islamic world has been a matter of local skirmishes by rebels (e.g. Babak) or warlords (e.g. Muhammad b. Musafir)[8] trying to weaken the central Islamic power. Neither side seems deeply interested in seeing the connection and essentially local characteristic of these interactions. Indeed, one can argue that the entire region between the Caspian and Black seas, and south of the Caucasus Range, should in fact be viewed as a single unit in political developments and social changes, despite its religious and linguistic differences and modern sense of identity.

Arguably, a glaring example of this is the case of the rebellion of Babak-i Khurramdin, the early ninth century "heretic" who took the region through a period of extreme turmoil and effectively changed the political landscape of it.[9] Studied within the context of Islamic history and heresiography, Babak and his rebellion seem as a last cry of Iranian, anti-Arab/Muslim elements hoping to "free" Iran from the Muslim yoke. This is a view taken up by both medieval Muslim historians and many modern Iranian historians, viewing the events from a nationalist point or view. The latter have surprisingly inspired an Azerbaijani nationalism which has turned Babak into a hero of the Turkic cause, this time, against the

---

[7] A brilliant modern example is Sebeos, *The Armenian History Attributed to Sebeos*, trans. Robert W. Thomson (Liverpool: Liverpool University Press, 1999), which benefits from detailed commentaries of James Howard-Johnston and Tim Greenwood, providing an excellent source for the study of the text of Sebeos in its contemporary context.

[8] For this see Kasravi, Part I.

[9] The definitive study of this rebellion, of course within its own focus, is now Patricia Crone, *The Nativist Prophets of Early Islamic Iran: Rural Revolt and Local Zoroastrianism* (Cambridge: Cambridge University Press, 2012). For a possible reflection of the Rebellion of Babak in the early Islamic Zoroastrian apocalyptic texts, see T. Daryaee, "A Historical Episode in the Zoroastrian Apocalyptic Tradition: the Romans, the Abbasids, and the Khorramdens," in T. Daryaee and M. Omidsalar eds., *Menog i Xrad: the Spirit of Wisdom: Essays in Memory of Ahmad Tafazzoli* (Costa Mesa, CA: Mazda Publishers, 2004), 64-76.

Iranians themselves!¹⁰

Within Armenian historiography, however, Babak himself is yet another semi-Muslim warlord, following on the footsteps of Ibn Ghanim or Muhammad b. Marwan, trying to alienate Armenians from their lands and setting himself up as a local magnate. Apart from assigning him a princess from the Siwni dynasty as a wife and mentioning the role of Sahl (Sahak?) Sambatean in his capture, Armenian history is not much interested in details of Babak's life and his actions in over two decades of him exerting power in the region. For Armenians, the most important aspect of Babak's rebellion is his bloody campaigns that resulted in much death.¹¹

As important and grave as the blood and the deaths were, however, one could argue that the importance of Babak cannot be reduced to local affairs and can also not be misconstrued as merely a nationalistic campaign of revenge set against the central Islamic rule. A cursory look at the entire episode would reveal that Babak's rebellion was not an isolated or unprecedented event in the history of southern Caucasus, and nor did its local influence vanish with the death of Babak himself. Viewed as part of the events of the late eighth and early ninth centuries, the Babak episode was an expected event, part of a regional need to fill the political vacuum left by a shift in south Caucasian politics and the results of greater shifts in the political centers of Islam. In the contemporary sense, it was also part of a local trend by the powers in the Caspian Sea region, particularly the Dailamites, to move further north and west and to overwhelm the southern Caucasus. This last goal, a major consequence of the rebellion, was indeed achieved by the establishment of the Mosafirid power, later in the ninth and tenth centuries, and the rising importance of the Dailamites as a whole in the medieval Islamic world.¹²

---

¹⁰ G. Riaux, "The Formative Years of Azerbaijani Nationalism in Post-Revolutionary Iran," *Central Asian Survey* 27, no. 1 (2008): 45-58. The Azeri case is certainly worth its own individual study and appears to rely on the view of Babak as an anti-tyranny rebel, perhaps similar to Robin Hood, with the original oppressor ("the Caliphs") being replaced with the new ones (the Iranian/Persian government).

¹¹ The Armenian accounts of Babak are best represented by the tenth century historian Movses Dasxuranc'i who was concerned with the history of Albania/Arran, Armenia's eastern neighbor, which bordered both Azerbaijan and Siwnik/Sisakan, the geographical spheres of Babak's influence; as well as Step'anos Orbelean, the thirteenth century bishop and historian of Siwnik'. Orbealean, more than anything, is writing a history of the Siwni family and is much concerned with placing them at a central position in the events he describes. Movses Dasxuranci, *The History of the Caucasian Albanians*, trans. C. Dowsett (Oxford: Oxford University Press, 1961). Step'anos Orbelean (Stephannos Orbelian), *Histoire de la Siounie*, trans. B. Brosset (St. Petersbourg: Eggers et Cie, 1865).

¹² The spread of Dailamites in the Near East and its similarity to the Norman expansion in Europe was first noticed and mentioned by Minorsky in the third part of his study of Caucasian history. See V. Minorsky, *Studies in Caucasian History: I. New Light on the Shaddadids of Ganja II. The Shaddadids of Ani III. Prehistory of Saladin* (Cambridge: Cambridge University Press, 1953).

The Rebellion of Babak-i Khurramdin

A basic outline of Babak's career and a consideration of its sources will provide us with the best point of departure. Here, I will highlight the highlights of Babak's life and rebellion and the related regional events that contributed to it, before proceeding to add more data that can illustrate the points about interregional context of the entire episode.[13]

Babak, allegedly born al-Hasan, and thus a Muslim, was born sometime in the last decade of the eighth century. Most sources, entirely Islamic in nature and thus quite hostile toward him, consider him to have been of very low-birth, with a father of non-local origins and possibly from al-Mada'in, the former capital city complex of the Sasanian Empire, at the time a suburb of Baghdad. Babak was probably born in Ardabil and, in his youth, joined initially the service of Muhammd b. Rawwad al-Shaddadi, the founder of the Arab Shaddadi dynasty of local rulers in Tabriz.[14] He is later associated with the village of Belālābād, near Meymand, as his home. It is also near this area that he comes to contact with Javidan (b.?) Shahrak,[15] the leader of the Khurramiya, a sect he later joins. His early carrier, aside from being in the service of local strongmen and dynasts, is presented as one of thuggery, involvement in violent and illegal acts, and the life of a brigand.[16] It is only after he joined Javidan's sect that he changed his ways, supposedly under Javidan's charismatic influence, and redirected his energy toward advancing the Khurrami cause. All these points, the low-birth, the foreign/non-local background, the lack of a career, and his quick adoption by a sect, are known narrative tropes provided to deprive him of a local context and clear purpose for his later rebellion.

Not too long after joining the Khurramis, Babak participated in a battle with a rival chief of the sect which resulted in victory for Shahrak's sect but also in his death. Babak, supported by Javidan's widow, soon rose to the position of the leader of the sect and, in 816, started a rebellion. This rebellion is immediately presented as one against the central Abbasid authority and was characterized by Babak's local campaigns and takeover of the city of Badhdh, just to the south of the Araxes/Aras River, establishing it as the base of his operations. The original impetus might have been a call to rebellion against Caliph

---

[13] A convenient summary can be found in Gh. H. Yusofi, "Bābak Ḵorrami," *Encyclopaedia Iranica Online*, 1988. The similar summary in Crone, 47-48, is uncharacteristically positivistic and uncritical of sources, including Waqid's book which has not survived in its original form and reaches us through accounts of Al-Maqdisi, Ibn Nadim, and eventually in Persian of Kaykawus b. Iskandar, *Qabusnama*.

[14] See Kasravi, Part III on this dynasty in Tabriz and later Ahar. The statement of Kaykawus b. Iskandar (*Qabusnama*, 195) that Babak had learnt to play the tanbur and used to sing to the public is among the many details that draws similarities between Babak and *Kur-Oghlu*, the Azerbaijani folk-hero who supposedly rose to rebellion during the Safavid period and similarly took refuge in a mountaintop fort.

[15] Ibn al-Athir, VI: 328. The patronymic, or perhaps second name, of Javidan is quite interesting. It is commonly written as سهرك s-h-r-k, with شهرك, Shahrak being a scholarly emendation as *Sahrak* is entirely meaningless. But could this be a simple Arabic take on Armenian Sahak/Isaac?

16 For a comparison of how different sources present Babak's character and actions. See Yusofi, "Bābak Ḵorrami."

Al-Ma'mun by Hatam b. Harthama, the "governor" of Armenia whose father was put to death by al-Ma'mun.[17] In either case, Babak seems to have gone his own way, mainly since Hatam died (or was killed?) almost immediately after starting his rebellion. In the next few years, Babak's power extended in the region, but was largely ignored as a local skirmish by al-Ma'mun who only sent some local forces and regional leaders to check his rising power.[18] According to al-Tabari, it was only on his deathbed that al-Ma'mun told his brother, Muhammad al-Mu'tasim, that he should curtail of Babak's rebellion. The latter, upon his elevation to the caliphal throne in 833, was faced with a full-scale military threat by Babak and soon dispatched Khydar Afshin,[19] his general and the local ruler of Ustrushana in Transoxiana, to destroy Babak.[20] After a two-year war, Afshin managed to eradicate Babak's power in the region. Babak, having fled his stronghold, evidently heading for Siwnik', was captured by Sahl b. Sambat (Sahl Smbatean) and was turned over to Afshin.[21] The rebel was taken to Baghdad where he was cut to pieces in front of al-Mu'tasim (AD 837). His brother and deputy, Abdullah, was also executed, this time at the hand of another Iranian nobleman from Dailaman. Before his death, the latter expressed happiness about being killed by a nobleman, someone of his own class and level, a rather strange expression for someone supposedly born of a travelling oil-salesman and a local prostitute.[22]

This story, of course, has several details that are of interest and can either make it suspect and open to revision or further clarification, or reveal points in the movement of Babak which have gone hitherto unnoticed. I shall first mention the points that are perhaps open to some speculation and historical investigation, before concentrating on those points which will hopefully help us to situate his rebellion more accurately.

The Name of Babak

Babak's name is given in all Islamic sources as Bābak, a name of unknown meaning and etymology. It was, of course, most famously held by the father of Ardashir I, the founder of the Sasanian dynasty (AD 224-241), where it is given in Middle Persian as *p'pky*. Armenian sources, most significantly Moses Dasxuranc'i in his History of Albania (Aghuank'), give it as a Bābān. In his *Muruj ul Dhahab*, al-Mas'udi tells us that Babak was born with the name al-Hasan[23] and that he evidently changed his name after joining the Khurramiyah sect.

---

[17] Ibid., 300.

[18] Muhammad b. Jarir al-Tabari, *The History (Vol. XXXIII: Storm and Stress along the Northern Frontiers of the 'Abbasid Caliphate)*, trans. C. E. Bosworth (Albany: CUNY Press, 1991), iii. 1100-1150.

[19] The name of Afshin, written either as حیدر or خیدر in the Arabic text, is in fact related to the "Hunnic" (Kidarite) name, Kidara. See K. Rezakhani. "A Note on the Alkhan Coin Type 39 and Its Legend," *DABIR* 1 (2015): 24-7.

[20] Tabari, *The History*, 1170 ff.

[21] Dasxuranc'i, *The History of the Caucasian Albanians*, III.20; Tabari, iii. 1223.

[22] Ibid., 1228.

[23] *Murūj* VII, p. 130.

However, it is worth noticing that al-Mas'udi, and presumably his source, is the only one mentioning this, and no others, including Dinawari or Ibn Nadim, provide such information.[24] On the other hand, it is quite strange for a rebel to choose the name Babak, a relatively unknown name with no religious or cultural association inspiring al-Hasan to "change" his name to Babak. Furthermore, the occurrence of the form *Baban* in Armenian leads us to believe that at least a degree of familiarity with the name existed among the Armenian writers. As unknown as Babak was in the contemporary Persian usage, it is rather common among the members of the house of Siwni, the traditional rulers of Siwnik'/Sisakan, the area bordering the lands of Aghuank and Azerbaijan. We know that the Siwni ruler who was a contemporary of Babak, Vasak III (780-821), had a grandfather called Babik/Babgen who was active earlier in the eighth century. It is perhaps no coincidence that Movses Dasxuranc'i (10th century) tells us that Babak married a daughter of this same Vasak III and was actively interfering in the affairs of Siwnik (see further below).[25]

Consequently, considering the uniqueness of the testimony of Mas'udi (the apostrophe should be the other way around) about the original name of Babak as *al-Hasan*, we might speculate that this testimony, much like Babak's disputed parentage, is a later addition and bears no contemporary support. Simultaneously, in relation to the house of Siwni, we might speculate that Babak's connections ran deeper than marriage and included a closer familial connection with the Siwnis, including bearing a name connected to that dynasty.

Babak's Parentage

It is well known that the Islamic sources on Babak's parentage are widely differing and unreliable, mostly designed to ridicule his descent and present him as a classic case of a base vagabond of low-birth. Perhaps the most telling evidence of this is the incredibly offensive way in which his mother is presented and ridiculed as a one-eyed woman who had illicit relationships with his father. To add insult, and perhaps ridicule the fallen rebel even more, a Persian source calls her Māhrū "Moon-face, beautiful", in contrast to her supposed missing eye.[26] Babak's father, in turn, is presented differently in all sources, except those who used *Akhbar al-Babak* of Waqid b. 'Amr al-Tamimi, including Ibn Nadim and al-Maqdisi (*al-Bad' wa-l-Tarikh*). He is called Mardas, Abdullah, Muṭahhar, Matar[27], or 'Amir b. Ahad, with his origins given as al-Sawad (lower Mesopotamia) or al-Mada'in (the Ctesiphon area, the former capital of the Sasanians). His profession is given as a merchant or a vagabond or a cooking-oil salesman, the latter, related by Waqid, being the most accepted among modern scholars. All these point to the low-birth of Babak, with a father immigrating to Azerbaijan

---

[24] Crone considers the name al-Hasan to be original and along with Mu'awiyya, Abdullah, and Ishaq, considers them common names which were picked up by Babak's parents because of their currency. Crone, *The Nativist Prophets*, 51.

[25] Dasxuranc'i, III. 19.

[26] Fasīh e Khwāfī, I.283.

[27] This might have been a brigand by that name who later claimed to have fathered Babak after sleeping with his mother, Mahru; Crone, 56.

from Mesopotamia and a local mother of ill-repute who, after the death of his father, became a wet-nurse in order to raise her sons, Babak and his brothers. Perhaps the only bright point in Babak's ancestry is mentioned by Dinawari who considers Mutahhar, the father of Babak, to have been a son of Fatima, the daughter of Abu Muslim al-Khurasani, the pro-Abbasid commander. Dinawari connects this Fatima, and the Fatimiyya in general, to Abu Muslim and counts Babak as the leader of the Fatimiyya sect of the Khurramis.[28]

However, as already noticed and pointed out by Gholamhossein Yusefi, two pieces of evidence present a different, more complementary parentage and ancestry for Babak. One is the letter of Babak to his son, quoted by al-Tabari, where he calls himself a king and a ruler and disowns his son who is of low birth.[29] Another is the evidence of an exchange between Babak's brother and deputy, Abdullah, and his designated executioner, Ibn Sherwin. Before submitting to his fate, Abdullah asks Ibn Sherwin who he is, and upon hearing that he is the king of Tabaristan,[30] Abdullah rejoices in the fact that he is being executed by someone of his own social rank and dignity.[31]

Both these testimonies, particularly the second one, point to the memory of Babak's family of a noble background, as well as their familiarity with Iranian nobility, particularly the surviving post-Sasanian aristocrats who found refuge in the mountains and valleys of the southern Caspian Sea. It is, in fact, quite remarkable that Abdullah is so intimately familiar with Ibn Sherwin's ancestry, and obviously not with him personally. One can conclude that Babak and his retinue had a sort of familiarity with the Caspian region and its political standing in the third century AH and understood the authority and the noble descent of its rulers. Additionally, they felt some manner of connection and relation, and equality in rank, with these princes who, in fact, claimed descent from the Sasanian noble families, such as the Karens or even the Sasanians themselves (as was the case of the Gāvbara family).[32] We could then argue that like the names and professions of his father and mother, which were obviously fabrications in order to tarnish Babak's reputation, his low-birth is also a story designed to re-enforce the same narrative. What his actual descent might be is something that can only be speculated, and even that purely based on possible contemporary political events and marginal narratives and notices, such as the one provided by Movses Dasxuranc'i and mentioned above. I will discuss some of these possibilities toward the end of the paper and after a discussion of the wider context of Babak's activities.

---

[28] Yusofi, "Babak Korrami."

[29] Tabari, 1221.

[30] This is probably Qarin (Karen) b. Shahriyar b. Sherwin, of the Bavandid dynasty, who was the contemporary ruler of Tabaristan. See Tabari, note 236.

[31] Ibid., 1231.

[32] For a history of these dynasties, see Ibn Isfandiar, *Tarikh-e Tabaristan*, ed. M. Mehrabadi (Tehran: Ahl-e Qalam, 1381). A basic introduction is given by W. Madelung, "The Minor Dynasties of Northern Iran," in *The Cambridge History of Iran, Volume 4: From the Arab Invasion to the Saljuqs*, ed. R. N. Frye (Cambridge: Cambridge University Press, 1975), 198–249

The Geographical Setting of Babak's Rebellion

Babak and his rebellion are intimately connected to his stronghold of Badhdh, a fortified mountain village in the region of Maymand. Apart from the fact that it is universally placed south of the Aras/Araxes river, however, we have no clear evidence of where it is located.[33] It was perhaps first Ahmad Kasravi who suggested a location in "Qaraja-Dagh, to the north of Ahar or slightly to the east" as the possible location of Badhdh.[34] This was the basis for which the Iranian Archeology Department identified the ruins of *Qala-ye Jomhur*, near the village of Kalibar, north of Ahar, as the site of Babak's fortress of Badhdh.

This identification, for Kasravi's purposes, put the region under the suzerainty of the house of Rawwad, whose history he was trying to narrate. Muhammad b. Rawwad, Babak's sometime employer, was a known Arab chief who fortified the site of Tabriz and made it into a major urban site.[35] In describing the history of his descendants, who come to rule the area of Ahar, but not Tabriz, a hundred years after Muhammad b. Shaddad himself, Kasravi is thus trying to justify the control of Ahar, and so sees the presence of Babak, a former employee of Muhammad b. Rawwad, as evidence that the area of Ahar was already in possession of Muhammad and that Babak's presence there was quite natural in this context. As a result, he puts the location of Badhdh also within this region. One can conclude that the Iranian Archeology Department's identification of the site, in this case, was then a tautology, particularly given Kasravi's intellectual influence.

This, on the other hand, presents a circular argument, since there is no actual evidence that Muhammad b. Rawwad had any power beyond Tabriz itself, as far east and north as Ahar or, indeed, further north, in Kalibar. Instead, our evidence is that the region entrusted to Rawwad, the father of Muhammad, was "Tabriz and its environs, including Badhdh," thus putting the latter much closer to Tabriz than Kasravi's suggestion, and perhaps to the north of it, on the present border of Armenia and Iran. What this means in the larger scale of Babak's revolt we will discuss further below.

The Rebellion of Babak in the Armenian Sources

The reputation of Babak as a violent rebel is a significant part of his lore in all sources mentioning him.[36] This is perhaps the most important aspect of the narrative as it is shared between Islamic and Armenian sources and, consequently, the root of the modern convention. While commenting extensively on the atrocities committed by the Khurramids, both set of sources stop short of providing a context or reason for it, beside Babak's personal

---

[33] Yusofi, "Badd," *Encyclopaedia Iranica*, http://www.iranicaonline.org/articles/badd

[34] Kasravi, 149. Laurent also locates Badhdh on the Aras River, north of Ardabil (obviously not directly north, which would be Shervan), on the border with Siwnik'/Sisakan. See J. Laurent, *L'Arménie entre Byzance et l'Islam depuis la conquête arabe jusqu'en 886*, ed. Maurius Canard (Lisbon: Librairie Bertrand, 1980), 161.

[35] Kasravi, 150.

[36] This is no doubt how Crone assumes it as well. Crone, 64-76. Also Daryaee 2004 who sees the reflection of this violence in the Zoroastrian apocalyptic texts.

cruelty. Here, I would like to suggest that the Armenian sources in particular have to be read against the background of local Armenian political competition, or lack thereof at this point, and the role of Babak in filling the gap left by the Battle of Bagravand and other events that created a political vacuum in early eighth century southern Caucasus.

Like the Battle of Avarayr (AD 451), a central point of Armenian elite identity, marking the rise of the Mamikonean house within the late antique Armenian history, the unfortunate Battle of Bagravand (25 April 775) has been raised by historians such as Łewond as a focal point of Armenian medieval history.[37] Scholars often consider Bagravand as a major watershed in Armenian history,[38] although this might be an exaggeration aimed at increasing the status of the Bagratuni family.[39] From an Armenian point of view, the Battle of Bagravand was the beginning of the process that resulted in the creation of the semi-independent Bagratuni kingdom of Armenia. From the point of view of the Islamic sources, however, the event and its aftermath went largely unnoticed, however, leaving little trace in the major works of history such as Baladhuri or al-Tabari.

Cyril Toumanoff, in several works dedicated to the history of mediaeval Caucasus, studied the immediate political and social aftermaths of Bagravand where he focused on Armenian sources.[40] Through his careful study of the sources, we have a clear idea of the competition between the mediaeval Bagratunis, the descendants of Ashot Msaker, and the Arcrunis of Vaspurakan and their struggle to gain supremacy over Armenian lands, as well as the foundation of the Iberian kingdom of the Bagration through the efforts of Adernase and his progeny.[41]

However, what is less explained is the role of Babak in these affairs and the impetus that his rebellion and its eventual defeat might have provided for the actions of Ashot Msaker and the eventual rise of Ashot I. It is worth pointing out that, between Bagrawand and the rise of Ashot the Great, there is a 100-year gap. In fact, before 775, the world of Azerbaijan and Arminiya is one dominated by the Islamic rule and administered through an extension of the

---

[37] This is based on the monumental work written shortly after the events, perhaps as early as 788. As expected, Łewond is also really writing a dynastic history, this time of the Bagratuni! Łewond, *History of Lewond, The Eminent Vardapet of the Armenians*, trans. Zaven Arzoumanian (Wynnewood: St. Sahag and St. Mesrob Armenian Church, 1982).

[38] Mark Whittow, *The Making of Byzantium, 600-1025* (Berkeley and Los Angeles: University of California Press, 1996), 213.

[39] Alison M. Vacca, "Language, Power, and Storytelling: Arabic in Caliphal Armenia," presentation at the conference *Navigating Language in the Early Islamic World* (Knoxville Tennessee, 7 April 2018). This is also well-argued in Alison Vacca's forthcoming new translation and commentary on the text of Łewond.

[40] Cyril Toumanoff, *Studies Christian Caucasian History* (Washington DC: Georgetown University Press, 1965), 563; 566.

[41] See a new take on this in Alison M. Vacca, "Conflict and Community in the Medieval Caucasus," *Al-ʿUṣūr Al-Wusṭā* 25 (2017): 66–112.

Sasanian administrative system.[42] After the middle of the ninth century and particularly after the death of al-Mutawakkil (AD 861), however, the nature of this world had changed, with the caliphal rule disappearing into a pale background and replaced by a series of Dailamite and Kurdish rulers competing and cooperating with strong Armenian kingdom over local power. Indeed, while before 775, the focus of Caucasian politics is toward Baghdad, in the post 850, the focus changes more toward the Caspian coast and the interior of Iran, a focus which comes to a dramatic conflict in 107, in the vicinity of Manzikert.

In Armenian sources, Babak is called Baban and is said to have come to Armenia with the invitation of Vasak Siwni and having married his daughter (whose name is not mentioned).[43] While Dasxuranc'i also mentions the same connection, he is more exact about the timing of Babak, putting him in the ninth century.[44] Step'anos Orbelean, however, connects Vasak's invitation of Babak to the campaigns of Marwan b. Muhammad (the later Umayyad caliph Marwan II) in 727 (Armenian year 176), so a hundred years earlier than the actual events. Both accounts, however, mention the murder of Varaz-Trdt, the Mihranid prince of Aghuank by the hand of Narseh P'ilipean of the Siwni. Babak is also blamed for the destruction and looting of the monastery of Mok'enots'. The murder of Varaz-Trdat of Aghuank, in both sources (Orbelean and Dasxuranc'i), is mentioned separately and evidently independently of the actions of Babak. However, Orbealean mentions that Babak and his ally, Aplasad (Abul-Asad?) of Aghuank were the ones who subjugated a major rebellion in Baghk which resulted in a massacre.

Orbelean does not mention the fate of Babak, preferring to detail the fate of the survivors of Mak'enots', while Dasxuranc'i correctly points to the role of *Sahl-i Sambatean* as the one responsible for the capture of Babak. Instead, due to his chronological problem, Orbelean makes a certain Sahak Siwni the contemporary of both *Sparapet* Smbat Bagratuni (perhaps Smbat VIII the Confessor) and the caliph al-Ma'mun. This Sahak is made to have joined forces with Smbat and his son-in-law, a man of "imperial" line, in a rebellion against the Muslim *ostikan*, Xul.[45] Sahak was killed during the battle, on the banks of the Hrazdan river, in AD 821, and was succeeded by his son, Grigor Sup'an, as the lord of Siwnik/Sisakan.

It is thus interesting that, in Orbelean's account, there was no rebellion of Babak during the time of al-Ma'mun or his successor, and Sahl-i Sambatean is not even mentioned in his account as the lord of Aghuank, the pride of place being given to Varaz-trdt and his descendants, including the brave maiden, Shahanduxt.[46] Vladimir Minorsky, in one of his most famous works, detailed the relationship between Babak and Sahl Smbatean, the local

---

[42] Alison M. Vacca, *Non-Muslim Provinces Under Early Islam: Islamic Rule and Iranian Legitimacy in Armenia and Caucasian Albania* (Cambridge: Cambridge University Press, 2017).

[43] Orbelean, 33.

[44] Dasxuranc'i, 19.

[45] Orbelean, 37.

[46] Orbelean, 36.

magnate, who finally betrayed Babak to the armies of the Caliph al-Mu'tassim.[47] Among the things that emerge is that Sahl is actually considered a local magnate of Arran/Aghuan and part of the Mihranid dynasty that ruled the area, although no relationship is mentioned between him and Varaz-Trdt or the aforementioned Aplsad.[48]

The multiplicity of characters and the importance of outsiders, such as *Sewada*, a son in law of Smbat VIII and his co-conspirator in the rebellion against "Xul",[49] shows the uncertain state of affairs in Armenia and Aghuank at the time. Various princes, survivors of Begravand, were trying to carve themselves areas of influence. While the affairs in Tayk or Vaspurakan, the territories of Arcrunius and Bagratunis, is much clearer, less is known about Aghuank or Siwnik, areas much closer to Iran and with borderland dynamics that made them susceptible to various influences. Naturally, one must be careful to not see the fate of Siwnik/Sisakan and Aghuank purely in terms of centralized Armenian affairs,[50] but within the larger context of the events taking place to their east and south, most importantly the regions of Aturpatgan and Dailam/Gilan. These were indeed where most of the future political fate of the region was shaped.[51]

Babak in the Islamic Sources

The main references to Babak in the Islamic sources consider his rebellion as a direct affront to the caliphal control of its frontiers in Aturpatgan and the Caucasus. Secondarily, the rebellion's religious nature and its closeness to the Mazdakism makes it an ideological threat to Islam and its authority.[52] The anti-Arab and anti-Islamic aspects of the rebellion also give it an Iranian, proto-nationalist nature, something that is mentioned in the primary sources and has been taken as forgranted in the modern characterization of his movement. Out of these, the first and the third aspects are of interest to us here; these can also point to the regional nature of the movement, the way it is characterized in the Armenian sources.

---

[47] Minorsky, *Caucasia IV*. On Sahl's betrayal of Babak, see al-Tabari,1228.

[48] Dasxuranc'i, 19.

[49] Orbelean, 37.

[50] It is perhaps necessary to point out that the consideration of the affairs of Sisakan/Siwnik' within that of Armenia is more a reflection of medieval Armenian historical tradition than one of reality. In fact, it is probably more prudent to consider the affairs of Siwnik' apart from the political competition in Armenia and see the activities of the Siwni family more in the context of Aghuank' and Azarbaijan. For a similar argument, see Tim Greenwood, *Sasanian Reflections in Armenian Sources* (Beverly Hills: Afshar Press, 2008).

[51] For the history of medieval Armenia, see René Grousset, *Histoire de l'Arménie: Des origines à 1071* (Lausanne: Payot, 1947).

[52] Crone, 279ff. A comparison with similar heretical movement, that of the Paulicians who were also accused by Armenians of conspiring with the Muslims, might be of further interest. See S. Dadoyan, *The Armenians in the Medieval Islamic World* (New Brunswick: The Transaction Press, 2011), 91-6. Also, Vrej Nersessian, *The Tondrakian Movement: Religious Movements in the Armenian Church from the Fourth to the Tenth Centuries* (Eugene: Pickwick, 1987).

We can perhaps best glance at this aspect from the point of view of the aftermath of the rebellion and the formation of local politics. Following the defeat of Babak in AD 836, through the efforts of Khydr Afshin, the control of the region as a whole falls to Abul-Saj Devdad b. Devband, a Sogdian deputy of Afshin.[53] The consequent levels of regional competition and power dynamics are quite interesting, with local strongmen such as Muhammad b. Rawwad or the Armenian naxarars trying to strengthen their position, while outsiders, such as Saweda, attempting to carve local power through marriage to the local dynasties and by brute force. Perhaps the most significant are Abul-Saj and his sons Muhmmad and Yusef who attempt to gain over-all supremacy, only to be suppressed by the still strong caliphal power. However, the vacuum eventually attracts elements from the neighboring regions, most importantly the Dailamites of Tarom, to become engaged in the local affairs. This is interpreted as part of the "Iranian Interlude" of medieval Islamic history and is connected to the rise of Dailamites, such as the Zayarids and the Buyids. As valid as this is, however, the rise of Muhammad b. Musafir and his son *Sallar* Marzuban appears to have had a much more local context.[54]

While the previous intruders from the south into Armenia had been representatives of the Islamic central state in one way or another, the Dailamites, coming from the east, did not represent the caliph in any form. We know that in the days of Babak, or in fact his brother Abdullah, there had been an attempt at using the Dailamite power in the Caucasian affairs, in form of creating a bond between the remnants of Babak's rebels and the Qarinid Mazyar (817-839),[55] but this had not actually been realized, nor can it be really qualified as a Dailamite involvement in the Caucasian affairs.

The Musafirids, or the Kankarid as they are alternatively known, were connected with marriage to the Justanids of Dailam. These were remnants of the local rulers of Dailam, dating back to the Sasanian period and in fact holding most of the power in the southern Caspian region, alongside related dynasties of Dailam and Tabaristan. The founder of the Mosafirids was Muhammad b. Mosafir, a Dailamite warlord who had married a daughter of Justan III, the Justanid ruler.[56] After defeating his kinsman, Ali b. Vahsudan, in 919 and taking over Qazvin, Muhammad was elevated to the position of a ruler, basing himself in the Dailamite high fortress of Shamiran to the south of Ardabil. From here he started his campaigns in Aturpatgan and the Caucasus. It is worth noting Muhammad's patronym here, as his father, called Mosafir "the traveler", indeed carries an unusual name. The normal speculation is that Mosafir is a "translation" of Persian Asvar/Asfar, a common name among the Dailamites (c.f. Asfar b. Shiruyeh).[57] This suggestion seems plausible, but we should also

---

[53] Tabari, 1222; also, Kasravi, Part 1.

[54] Ibid., Part 1.

[55] Tabari, 1269; Crone, 66-8.

[56] C. E. Bosworth, "Mosaferids," *Encyclopaedia Iranica Online*, 2013,

http://www.iranicaonline.org/articles/mosaferids/

[57] Kasravi, 35.

notice that the last of the ruling Sajids, whom the Mosafirids really replace as the major Muslim power in eastern Armenia/Aghuan, carried the *kunya* of Abu-Musafir.[58] Could this name by any chance connect the Musafirids to the Sajids, as names often indicated family relations in this period? Muhammad's own sons, Vahsudan and Marzuban, carried clearly "Justanid" names, which show their connection to that dynasty. So, we might need to think about Kasravi's suggestion twice before dismissing the name of Muhammad's father as a pure Arabisation.

Muhammad's arrival in the Caucasus has been remarked on memorably by Movses Dasxuranc'i in his *History of the Caucasian Albanians* in this manner: "Now when these times had passed [times of the Sajids] and when the Arab people had become exhausted, yet another people appeared. They were called Dailamites. Their chief was a certain man named Salar, and he widely extended his authority and came to rule over the Aghuans, Persians, and Armenians. He came to Partaw and made it his immediately."[59] This Aghuanuan branch is, in fact, the branch of the Musafirids that Ahmad Kasravi calls the Sallarids, to distinguish them from a branch that continued to rule in Tarom.[60]

The rise of the Mosafirids from the Caspian region, and indeed the vicinity of Ardabil, is quite comparable to the rise of Babak who similarly rose from the Ardabil region and eventually got involved in the affairs of southern Caucasus, Aghuank and Siwnik'. Much like Babak, they came from an unknown background, having married into the more noble family of the Justanids. They also represented a minority religion, that of Zaydite Shi'ism, but were representatives of the hostile Muslim enemies of the Armenians. The fact that most of their efforts was concentrated on the regions in the east both shows a similarity of the patterns of pollical dominance, as well as the closeness of the politics of Aghuank and Siwnik' to that of Dilaman and other southern Caspian regions.

Comparing the rise of the Musafirids with that of Babak and considering the particular details in the story of Babak mentioned above, we might be able to suggest some possible solutions to these problems. As unknown as Babak's name and parentage are, we have provided clues and possibilities about their local contexts above. Further considering the origins of Babak from the Ardabil region and his claim to a noble parentage, we might suggest that he indeed belonged to a local noble family, perhaps similar to Muhammad b. Musafir, who enjoyed a sort of local influence and closeness to the Dailamite rulers. This could justify both Babak and Abdullah's claim to noble background, as well as the impressive knowledge of Abdullah about the ancestry of Ibn Sharvin, his executioner. Further, the occurrence of the name of Babak, indeed almost a unique name, among the members of the Siwni dynasty, provides another local connection, perhaps to the nobility of the southwestern Caspian region, those of Shervan, Aghuank', and Siwnik'.

---

[58] Madelung, 230.

[59] Dasxuranc'i, 21.

[60] Kasravi, 39.

Conclusion

The Rebellion of Babak is a historical episode that can be viewed from the point of view different sources. Because of its geographical placement on the periphery of the Islamic and Iranian worlds and in the neighborhood of rising medieval polities in the Caucasus and the Caspian region, it can also be considered from a regional and extra-regional viewpoints. Traditionally, this episode has been seen by modern scholars as a watershed of Islamic and Iranian history, either marking a strong show of force by the Abbasid Caliphate, or exemplifying a nativist rebellion against Islamic rule by the Iranian/non-Arab elements. This latter case has then given rise to the coopting of Babak and his movement by regional identities such as Azarbaijan, where he has been raised as a national hero.

The consideration of the above points, pending more in-depth research in the primary sources and perhaps acting as a new window through which to view them, can allow us to see the rebellion of Babak as neither a local, isolated event undertaken by a total outsider nor as an organized, nationalistic "Iranian" struggle against "oppressive" "Arab" forces. Instead, by placing Babak against the background the local dynasties and geographical settings, we could see his rebellion as an expected outcome of local political events and his actions in line and consistent with the political competitions that eventually resulted in the formation of medieval southern Caucasus and Azarbaijan. Through this consideration, we would be able to draw a picture of the events surrounding Babak's movement as being part of a long-term trend in the region that results in the gradual formation of local polities in the area, as well as a growing connection to the Caspian region, resulting in the rise of dynasties such as the Musafirids later in the ninth and tenth centuries. From an identity point of view, placing Babak within Azarbaijani, Dailamite, Armenian, and perhaps an Aghuank and Siwni contexts would show the futility of attempts at defining his identity and that of his movement from a modern point of view and the need to avoid anachronistic labels and identities.

AMBIGUA GENS? METHODOLOGICAL PROBLEMS IN ANCIENT ARMENIAN HISTORY

Giusto Traina

When scholars of Antiquity approach a "marginal" state such as the Kingdom of Greater Armenia (*Mec Hayk'*), they somehow find themselves in the same situation as the pioneers of modern ethnography. A good example is the British traveler Mary Edith Durham who wrote her accounts on the Western Balkans at the beginning of the twentieth century. Durham eventually realized that, while she was incapable of seeing these territories with Eastern eyes, she was equally incapable of seeing them with Western ones.[1]

In fact, historians of antiquity tend to evoke Pre-Christian Armenia only in the context of a military crisis. This is mostly due to Armenia's strategic position between East and West or, more precisely, *in* East and West, as well as to the considerable extension of its territory, and the wealth of its sites and natural resources. Hence, the endless war between Rome and Iran to control the kingdom[2]. In a statement, which became rather popular among political scientists, Edward Luttwak said that "only Armenia was a true buffer state, serving as a physical neutral zone between the greater powers of Rome and Parthia, and providing them with a device that would serve to avoid conflict as long as they desired to avoid conflict."[3]

Yet, if we want to examine the history of Armenia from the mid-first century BCE and the third century CE, utilizing military analysis comes short. Armenia's contacts with Rome and Parthia gave it a mixed identity. This renders its study more difficult as both Classicists and Orientalists hardly communicate with scholars of Armenia, subsequently leading to

---

[1] M.E. Durham, *The Burden of the Balkans* (London: Edward Arnold, 1905), 287; M. Todorova, *Imagining the Balkans*, Updated Edition (Oxford: Oxford University Press, 2009), 121.

[2] R. M. Sheldon, *Rome's Wars in Parthia. Blood in the Sand* (London-Portland: Valentine Mitchell, 2010).

[3] E. N. Luttwak, *The Grand Strategy of the Roman Empire from the First Century AD to the Third* (Baltimore: Johns Hopkins, 1976), 24.

Armenia holding a marginal place in the historical continuum. Ancient historians must cope with an uneven evidence: for a Classicist it is extremely difficult to analyze early Medieval Armenian sources that handle Pre-Christian history from the vantage point of the fifth and sixth centuries.[4]

Moreover, the most important local source of pre-Christian Armenia, Movsēs Xorenacʻi's *History of Armenia* (*Patmutʻiwn Hayocʻ*), is also the most controversial one on several counts, including its datation. As Xorenacʻi mixes local oral traditions with heterogeneous Greek and Roman sources, often confusing or telescoping the rulers and the dynasties, the final result is a chronological hodgepodge.[5] Classical scholars virtually dodge Xorenacʻi, resting on Robert Thomson's introduction of his valuable English translation (1978), where the Armenian "Father of History" is depicted more or less as a mischievous forger.[6] Thomson's conclusions, which are not supported by strong historical arguments, provided classical scholars with a good reason to dismiss Xorenacʻi's work.[7] For an ancient historian, the "Western" Greco-Roman evidence is easier to handle. Classical sources essentially focus on the policy of the kings, who mostly tried to hold a geopolitical role in the balance of power between Rome and Iran (the only exception was the ephemeral empire of Tigran the Great (95-ca.55 BCE), who was strong enough to disavow its status of "friendly king").[8]

I will give some examples:

a) In the brief description of Roman power, at the end of the sixth book of his *Geography*, Strabo states that the Asiatic regions of the geopolitical sector between the Black Sea and the Caspian were ruled by "subject kings". This was also the case of the three

---

[4] T.W. Thomson, *Moses Khorenatsʻi. History of the Armenians, Second Edition* (Ann Arbor: Caravan Books, 2006); G. Traina, "Tradition et innovation dans la première historiographie arménienne," in *L'historiographie tardo-antique et la transmission des savoirs*, eds. Ph. Blaudeau and P. Van Nuffelen (Berlin, New York: De Gruyter, 2015), 153–164.

[5] Sargsyan tried to justify Xorenacʻi's chronological system, with little success. G.X. Sargsyan, *Hellenistakan tarašrǰani Hayastanə ev Movses Xorenacʻin* [The Armenia of the Hellenistic period and Movsēs Xorenacʻi] (Erevan: Academy of Sciences, 1966). See also L. E. Patterson, "Caracalla's Armenia," *Syllecta Classica* 24 (2013): 73-199, 179.

[6] Thomson, 1-60.

[7] See E. Kettenhofen, *Tirdād und die Inschrift von Paikuli* (Wiesbaden: Reichert, 1995); G. Traina, "Review of Kettenhofen," *Mesopotamia* 31 (1996): 308-11; E. Kettenhofen, "Römische Kaiser des 3. Jhs. n. Chr. in den Patmutiwn Hayocʻ des Movsēs Ḥorenaçi", in *Diwan. Untersuchungen zu Geschichte und Kultur des Nahen Ostens und des östlichen Mittelmeerraumes im Altertum. Festschrift für Josef Wiesehöfer zum 65. Geburtstag*, ed. C. Binder, H. Börm, A. Luther (Duisburg: Wellem Verlag, 2016), 665-682; or, more radically, M. Heil, *Die orientalische Aussenpolitik des Kaisers Nero* (München: Tuduv, 1996), 12 f.; G. Kreucher, *Der Kaiser Marcus Aurelius Probus und seine Zeit* (Wiesbaden: Franz Steiner, 2006), 39; 158f.

[8] The most thorough study is still from M.-L. Chaumont, "L'Arménie entre Rome et l'Iran I. De l'avènement d'Auguste à l'avènement de Dioclétien," *Aufstieg und Niedergang der Römischen Welt* 9.1 (Berlin-New York: De Gruyter, 1976), 71-194.

Subcaucasian kingdoms of Armenia, Albania, and Iberia, which

need only the presence of leaders and act well when ruled, but they revolt when the Romans are otherwise occupied. This is also the case with those beyond the Istros around the Euxeinos, except for those living around the Bosporos and the nomads, for the former are subjects but the latter, because of their great unsociability with everyone, are useless and need only be watched.[9]

Armenia owes its reputation from its large dimension. In his conception of the *oikoumene*, Strabo depicts Armenia as a transitory mountain zone, between the Mediterranean and the steppes.

b) Hinting at the same historical context, Tacitus labels Armenia in more "imperialist" terms:

[The Armenians] have been an ambiguous race from ancient times, both in the instincts of the people and in their country's situation, since, extending a broad frontier along our provinces, they stretch deep into the Medes [=the Iranians] they are interposed between, and more often disaffected toward, these greatest of empires, with hatred for the Romans and resentment of the Parthian."[10]

The key expression here is *ambigua gens*, "ambiguous race." Further on, when he narrates the outbreak of the Roman-Parthian war in 58 CE, Tacitus speaks of the "ambiguous loyalty" of the Armenians (*Armenii ambigua fide utraque arma invitabant*),[11] portraying them using the cliché of the untrustworthy Oriental. According to the Roman conception of *bellum iustum*, the "Armenian ambiguity" represented a good pretext to justify invasion. However, the expression *ambigua gens* was not a mere chauvinistic jeer. A keen observer of geopolitics and a master of style, Tacitus handled the multiple nuances of the adjective *ambiguus*. Although it can be translated as "untrustworthy" or "equivocal," its principal meaning is "uncertain," "vague," "undecided". Therefore, the concluding sentence of *Annals* 2.56.1, where Tacitus points at the Armenian disaffection towards the Roman and the Parthian superpowers, seems to give the Armenians a sympathetic ear. Like in the famous speech of the *Life of Agricola*, where the Caledonian chieftain Calgacus accuses the Romans of making a desert and calling it peace,[12] Tacitus seems, if not to justify, at least to understand

---

[9] Strabo, *Geography*, 6.4.3, tr. D.W. Roller, *The Geography of Strabo: An English Translation, with Introduction and Notes* (Cambridge and New York: Cambridge University Press, 2014). On Strabo's Armenia, see G. Traina, "Strabo and the History of Armenia," in *The Routledge Companion to Strabo*, ed. D. Dueck (London and New York: Routledge, 2017), 93-101.

[10] Tacitus, *Annals*, 2.56.2, trans. A.J. Woodman, *Tacitus, The Annals. Translated with Introduction and Notes* (Indianapolis and Cambridge: Hackett, 2004).

[11] Ibid., 13.34.2. On ambiguity in Tacitus, see J. Hellegouarc'h, "Tacite et l'ambiguïté," in *Hommages à Carl Deroux. 2. Prose et linguistique, médécine*, ed. P. Defosse (Bruxelles: Latomus, 2002), 216-23.

[12] Tacitus, *Agricola*, 30.5.

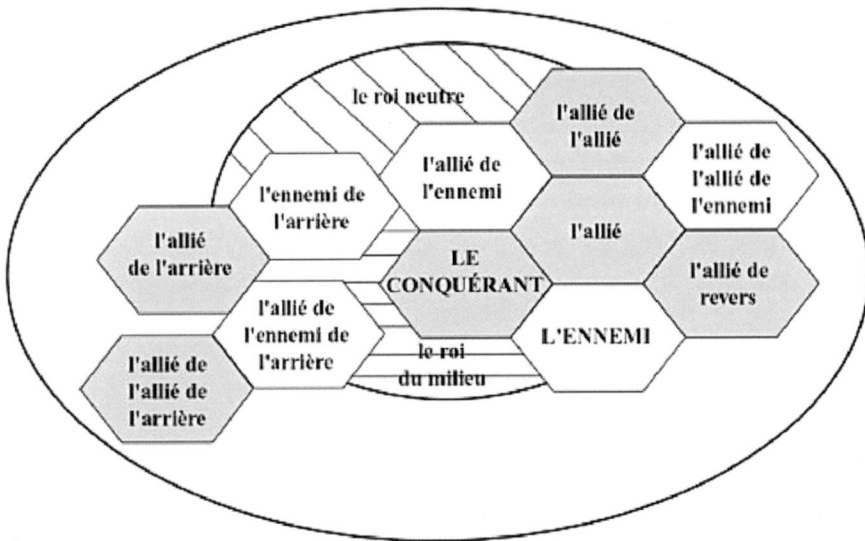

Figure 1. The "circle of kings" according to the *Arthaśāstra* (drawing by Bernard Mangin, from Chaliand 2006)

the reasons of the "ambiguous" character of the Armenians: they were untrustworthy and undecided because they could not just choose between the Romans and the Iranians, who both desired to control the country.

Maybe we need some change of perspective here. A helpful source would be the *Arthaśāstra*, a Sanskrit treatise on statecraft elaborated between the first century BCE and ca. 300 CE and attributed to Kauṭilya.[13] This astounding work developed the complex theory of the *maṇḍala*, "the circle of kings," according to which a conqueror king must not only consider his enemy but also the minor kings implied in the strategic balance and their degree of loyalty or friendship. An important role was played by the king called *madhyama*, "intermediate": "One with a territory immediately contiguous to both the enemy and the seeker after conquest, and who is able to assist them both when they are united and they are not and to overpower them when they are not united is the intermediate."[14] As Patrick Olivelle observed, "the only two kings Kauṭilya considers outside the *maṇḍala* theory of ally and enemy are the *madhyama*, who is an intermediate king located between two enemies, and the *udāsīna*, a powerful king who remains, or can afford to remain, neutral."[15] Gérard Chaliand highlighted the importance of this treatise for modern strategic theories (Fig. 1).[16]

---

[13] For date and authorship, see P. Olivelle, *King, Governance, and Law in Ancient India: Kautilya's Arthasastra* (Oxford: Oxford University Press, 2013), 25-38.

[14] *Arthaśāstra* 6.2.21, tr. Olivelle 2013. See L. Künhardt, "Staatsordnung und Macht in indischer Perspektive: Chanakya Kautilya als Klassiker der politischen Ideengeschichte," *Historische Zeitschrift* 247 (1988): 333-55.

[15] Olivelle, 48.

[16] G. Chaliand, "Kautilya et la naissance du politique," in *Arthasastra. Traité du politique*, nouvelle édition (Paris: Pocket, 2016).

Figure 2. Detail of the Great Cameo of France (Paris, Bibliothèque Nationale de France)

The whole political history of Armenia may be studied from this point of view.

Although refined and useful for modern theories of strategy and management, this political doctrine neglects a major factor, a capital one for Armenia: a foreign power willing to secure control of the kingdom had to gain the favor of its ruler by making a *madhyama* of him.

c) Visual evidence needs to be examined too. Possibly, Armenia's subjection may be detected in the "Grand Cameo of France," the large sardonyx presenting a complex iconographical program illustrating the continuity of the Julio-Claudian dynasty (Fig. 2). [17]
The central register is dominated by Germanicus, standing in front of Tiberius and Livia, herself a supporter of a strong Oriental policy. By her throne, an Oriental figure seated with a sad and resigned attitude could well be a personification of Armenia.

d) We find another visual example in the panel of the temple of the imperial cult in Aphrodisias, the *Sebasteion*, showing Nero and the personification of Armenia (confirmed by the

---

[17] According to others, the figure might be Parthia. See L. Giuliani, *Ein Geschenk für den Kaiser. Das Geheimnis des Großen Kameo* (München: Beck, 2010), 26. The dating of the cameo is controversial. Giuliani supports a dating around 23-24 CE. Giuliani, 44; see Giuliani, 45-61, for criticism on former hypotheses). But the years 14-18 cannot be excluded, as argued by S. Mazzarino, *L'impero romano*, Second edition (Roma and Bari: Laterza, 1973), 855-68.

Figure 3. Nero and Armenia. Relief from the *Sebasteion* at Aphrodisias.

Greek inscriptions on the panel).[18]

This dramatic image hints at the iconography of Achilles and Penthesilea, represented on another relief of the same complex. It did not overtly imply wartime sexual violence, as suggested by Dick Whittaker, although its sophisticated message could be also thought at first degree, showing a weak and submissive Armenia (Fig. 3).[19]

Meanwhile, the situation had radically changed. The dynasty founded by Artašēs was history. After the treaty between emperor Nero and the Parthian king, Walagaš (*Vologases*) in 63 CE, a king of the Arsacid family, Trdat (*Tiridates*), was put on the Armenian throne. The dynastic change was the result of a compromise: after several attempts to impose a "friendly" king, Rome was forced to accept an Armenian king of Parthian descent, who was already ruling Armenia after the Parthian invasion of 52. But the Roman ambassadors managed to impose the Parthians a limiting clause: the new sovereign should accept Roman authority. Therefore, Trdat should go to Rome and receive the diadem by Nero himself. This compromise ensured peace for some years, until Trajan invaded the country in 114. The pretext was the deposition of a king named Exedares, or Axidares, by another Arsacid prince, Parthamasiris. Apparently, Exedares had not "been satisfactory neither to the Romans nor to the Parthians" (ὡς οὐκ ἐπιτήδειον οὔτε τοῖς Ῥωμαίοις οὔτε τοῖς Πάρθοις).[20] We ignore whether he was simply a bad king or did not show enough allegiance and affection to the other powers. Despite his usual blurred chronology, Movsēs Xorenac'i states that "when tumult and confusion arouse in the west, Artašēs took courage

---

[18] *Supplementum Epigraphicum Graecum* 31.920 a-b; 921.

[19] C.R. Whittaker, *Rome and Its Frontiers: the Dynamics of Empire* (London and New York: Routledge, 2004), 115-21; D. J. Mattingly, *Imperialism, Power, and Identity: Experiencing the Roman Empire* (Princeton: Princeton University Press, 2011), 100; P. J. Kosmin, *The Land of the Elephant Kings: Space, Territory, and Ideology in the Seleucid Empire* (Cambridge: Harvard University Press, 2014), 138. See also criticisms by C. Vout, *Power and Eroticism in Imperial Rome* (Cambridge: Cambridge University Press, 2007), 25-27; 48, n. 118.

[20] Cassius Dio 68.17.3, tr. E. Cary, *Dio's Roman History*, vol. iii, the Loeb Classical Library (London: Heinemann, 1925).

from these events to rebel against the Roman empire, withholding the tribute."[21] We see how Tacitus' statement about the Armenians can be applied to his own age, when emperor Trajan was preparing his expedition against Armenia and Parthia (114-117 CE).

The Armenian ambiguity can be also detected on a local scale. In the sixth century, many decades after the loss of their political independence, the identity of the Armenians mattered more than loyalty. Procopius of Caesarea, a rather chauvinistic historian, implicitly criticized the inhabitants of the region of Xorjayn, in Western Armenia, who seemed to ignore the very existence of a frontier between the Byzantines and the Sassanians: "...the inhabitants of this region, whether subjects of the Romans or of the Persians, have no fear of each other, nor do they give one another any occasion to apprehend an attack, but they even intermarry and hold a common market for their produce and together share the labors of farming. And if the commanders on either side ever make an expedition against the others, when they are ordered to do so by their sovereign, they always find their neighbors unprotected"[22].

The nature of the classical evidence gave way to the paradigmatic idea of Armenia as the bone of contention between East and West. The image of the Roman empire, or of the Roman imperialism, has long been a mirror of modern imperialist views of the East. For this very reason, Western textbooks of ancient history put Armenia in a more marginal position than it actually held. Apart of some episodes of Hellenistic history, and of the hints at the "endless Armenian question" within the narration of Roman imperial military history, Caucasian history is scarcely embedded in the continuum of ancient history. Although the new trends in Roman history are now focusing at a "decolonized" image of the empire, nonetheless it is difficult to process the later Armenian sources and consider them on the same level of the Greco-Roman ones. The historians of Rome have scarcely been attracted by the inner structures of the Armenian society.

In fact, and the political theories of the "Age of Empires" influenced the paradigms of the Roman history. For Lord Curzon, Armenia was no more than a Roman protectorate: among the "barrier of protected States" preventing direct conflict with Iran, it was "the most important of all, by reason of its physical features and geographical position... having a career which in its stormy vicissitudes has recalled to many writers the chequered and fateful experience, between the rival Powers of Great Britain and Russia, of the buffer kingdom of Afghanistan"[23]. More sympathetically, in the "Historical Summary" of Armenian history, attached to the "Blue Book" denouncing the genocide of 1915, a young Arnold Toynbee wrote: "was Armenia to be wrested away", he asked, "altogether from Oriental influences and rallied to the European world, or was it to sink back into being a spiritual and

---

[21] Movsēs Xorenac'i 2.54, tr. Thomson (see above, n. 4)

[22] Procopius, *Buildings* 3.3.10-12, trans. H.B. Dewing, *Procopius*, vol. vii, vol. iii, the Loeb Classical Library (London: Heinemann, 1940).

[23] G. N. Curzon, *Frontier* (Oxford: Clarendon Press, 1907), 38-9.

political appanage of Iran?"[24]. Although sympathetic, Toynbee was exercising "the positional superiority of the observer over the observed."[25]

Another popular cliché is to depict Armenia as a buffer state. I have already mentioned Luttwak's popular statement. However, David Braund showed that the buffer model is most inadequate: "just as there is a sense in which the kingdom was a buffer for Rome, so in a sense Rome was a buffer for the king and his kingdom."[26] In fact, in modern political science, a buffer state is a "small independent state lying between two larger, usually rival, states (or blocs of states)."[27] Armenia was independent. Except for some interruptions, the ambiguous kingdom *par excellence* maintained its independence for about six hundred and fifty years. But it was certainly no small state. It held a strategic position, a vast expanse of territory, and valuable natural resources. To neglect its role, or to diminish its position in the balance of power, is a mistake, as Armenia cannot be compared with invented buffer states like Afghanistan. With all due caution, the history of Greater Armenia may be compared with the history of Poland, which was not established as such but became later a buffer state, in the sense of a "state geographically located between two other states engaged in a rivalry," with an important function in the balance of power.[28] The failure to balance brought the death of the Armenian state in 428. We may use for Greater Armenia Tanisha Fazal's argument about the fall of Poland in 1795, "Poland's failure to balance… appears to confirm two conventional wisdoms: first, that Poland died because it was unable to govern itself; and second, that states that fail to balance will, indeed, be selected out of the system."[29]

To sum up, ancient historians could take advantage of a recent statement by Jo Laycock, who says that "Armenia does not fit easily into the dichotomous categories of 'East' and 'West' emphasized by [Edward] Said. Instead images of Armenia have been characterized by ambiguity and fluidity."[30] Laycock draws on the experience of post-colonial studies to frame Armenia in a more refined way. As a modern historian, she can rely on a more balanced dossier of sources. But a fair solution could well consist in considering ancient Armenia as

---

[24] A. J. Toynbee (ed.), *The Treatment of Armenians in the Ottoman Empire, 1915-16: Documents Presented to Viscount Grey of Fallodon, Secretary of State for Foreign Affairs, by Viscount Bryce, With a Preface by Viscount Bryce* (London: Sir Joseph Causton and sons, 1916), 601. See C. Mouradian, "Préface," in A. J. Toynbee, *Les massacres des Arméniens. Le meurtre d'une nation (1915-1916)* (Paris: Payot, 2004), 7-27.

[25] S. Subrahmanyam, "Connected Histories: Notes towards a Reconfiguration of Early Modern Eurasia," *Modern Asian Studies* 31 (1997): 735-62, 761.

[26] D. Braund, *Rome and the Friendly King. The Character of the King Clientship* (London-Canberra: Croom Helm, 1984), 93.

[27] T. Mathisen, *The Functions of Small States in the Strategies of the Great Powers* (Oslo: Universitetsforlaget, 1971), 107.

[28] T. M. Fazal, *State Death. The Politics and Geography of Conquest, Occupation, and Annexation* (Princeton: Princeton University Press, 2007), 39-40; 70.

[29] Ibid., 116.

[30] J. Laycock, *Imagining Armenia: Orientalism, Ambiguity and Intervention* (Manchester: Manchester University Press, 2009), 11.

a connected history. Armenia was not only between East and West, but also *in* East and West. Moreover, it was in the Caucasus. Twenty years ago, in a seminal article, Sanjay Subrahmanyam criticized nationalism for blinding us "to the possibility of connection."[31] This could be a useful step to break the current paradigms, no matter how they are resting on Romanocentrism, nationalism, or parochialism.[32]

---

[31] Subrahmanyam, "Connected histories", see above, note 25.

[32] V. A. Shnirelman, *The Value of the Past: Myths, Identity and Politics in Transcaucasia* (Osaka: National Museum of Ethnology, 2001); G. Traina, "Mythes fondateurs et lieux de mémoire de l'Arménie pré-chrétienne (i)," *Iran & the Caucasus* 8 (2004): 169-81.

EARLY MODERN

THE "GREAT SCHISM" OF 1773: VENICE AND THE FOUNDING OF THE ARMENIAN COMMUNITY IN TRIESTE[*]

Sebouh David Aslanian

In the 1901 bicentenary issue of the Mkhit'arist Congregation's flagship journal *Bazmavep* [Polyhistory], Father Step'an Sarian makes the following obscure allusions to events that shook the island of San Lazzaro during the second half of the eighteenth century, and forever altered its history. He writes, "Thus, dark clouds passed over the arches of San Lazzaro and internal dissent blackened, for a while, its luminous horizons."[1] This exceedingly cryptic and all too fleeting passage is a veiled reference to what may be called the Great Schism of San Lazzaro, an event that gave rise to a splinter order of the Congregation, first in Trieste in 1775, followed, in 1811, by Vienna, when the Congregation was cleaved into two different and often rival, bitterly factional orders. Patriarch Maghakia Ormanian is not alone when he notes the "certain cautiousness" with which Father Sarian, and nearly everyone

---

[*] An earlier version of this essay was presented as a paper at the "Confessionalization and Reform: The Mkhit'arist Enterprize from Constantinople to Venice, Trieste, and Beyond" conference at UCLA on December 15, 2017. I am thankful to Houri Berberian, Cesare Santus, Merujan Karapetian, Marc Nichanian, Stefania Tutino, Jesse Arlen, Hagop Gulludjian, Bedross Der Mattossian, and Gerard Libaridian for helpful comments. My transcriptions and translations from Italian have benefitted from advice and help provided by Francesca Ricciardelli, Cesare Santus, Zara Pogossian, and Stefania Tutino. All translations unless otherwise specified are my own as are all remaining errors of judgment in reaching my own conclusions. The title of the paper (the Great Schism) should not be confused with the earlier "Le grand schisme" of either the separation of the Orthodox Church from that of Rome in 1054 or the division of the Armenian Church from that of its neighbors during the sixth century (See Nina Garsoian, *L'Église arménienne et le grand schisme d'Orient* (Louvain: Peeters, 1999). I have chosen the adjective "Great" to distinguish this internal separation within the ranks of the Mkhitarist Congregation from several smaller and lesser significant schisms in the 1750s and the middle of the nineteenth century that, unlike the Great Schism of 1773, did not result in a separate order.

[1] Step'an Sarian, "Aknark mě Mkhit'arian ukhtin venetko ěndhanrakan abbayits' vra," *Bazmavep*, 1901, 51. "Այսպէս անցան Ս. Ղազարի կամարին վրայէն սեւաթոյր ամպեր. ներքին տարաձայնութիւններ առ ժամանակ մի խաւարեցուցին անոր պայծառ հորիզոնը։"

else who has followed him, has chosen to gloss over the nature of the disagreements and circumstances that led to the division of the order.[2] To the extent that they are even familiar with the Congregation's history, most scholars of Mkhit'arist history have a very imprecise understanding of this schism. At most, they might be aware, as the chronicler cited above, that, indeed, disagreements occurred on San Lazzaro in 1773, leading to the creation of a rival order in Trieste. If they have done their homework, they might even know that this order was later transferred to Vienna, where the congregation subsides to this day. However, their knowledge is limited to these basic facts, and this is largely because, over the last two centuries or so, despite the publication of numerous studies on the Congregation, nearly all by Mkhit'arists, a miasma of silence has fallen on this important chapter of the island's history.

Relying on previously unconsulted documents stored in the state archives of Venice, Trieste, Rome, and Vienna, as well as other evidence, much of which has remained either unknown or marginal to discussions on the history of the Mkhit'arists, this essay provides a preliminary outline of the history of the Great Schism of 1773. On the basis of a careful assessment of the available evidence, the essay argues that disputes over constitutionalism and representative governance in a monastic setting, and *not* theological or sectarian differences as Ormanian suggested a hundred years ago, were the leading factors that gave rise to the Great Schism. To this end, the essay begins with a brief overview of the life of Abbot Mkhit'ar and the history of his order in Venice, until his demise in 1749. This is followed by a detailed account of Step'annos Melkonian's election, in 1750, and the circumstances leading to the Great Schism, a little over two decades later. My discussion here situates the origin of the schism in the context of a domestic spying agency in Venice known as the *Inquisitori di Stato*, whose shadowy tribunal of three judges played a pivotal role in creating the schism and subsequently storing hundreds of pages of reports on the event in their archives. The third section sheds light on the creation of the Armenian community in the Habsburg port city of Trieste as a direct consequence of the Great Schism and briefly discusses the privileges granted to the Trieste branch of the Mkhit'arists by Empress Maria Theresa in 1775. In the context of studying this little-known 1775 decree, I provide a fresh look at the obscure history of Trieste's tiny Armenian community during the last quarter of the eighteenth century and argue that, much to the chagrin of the Venetian authorities, the expelled monks from Venice were welcomed with open arms by the Habsburg authorities

---

[2] Maghakia Ormanian, *Azgapatum: Hay ughghap'aṛ ekeghets'vo antsk'erě zkizbēn minchev mer orerě harakits azgayin paraganerov patmuats* [National History: The events of the Armenian Apostolic Church from the beginning until our days narrated alongside their national circumstances] (Istanbul 1914), 2: 3076-3084 (3077). "այլ թէ ինչ էին այդ տարաձայնութիւնները, բացատրել չեն սիրել, անշուշտ տեսակ մը զգուշաւորութեամբ:" Alexander Yerits'eants' in his nineteenth century account of the order's history was among the first to note the silent treatment that the schism received even a hundred years after its occurence. "...բայց ահա այս կէտը միչեւ այսօր էլ լռութեամբ են անցնում թէ Վենետիկի եւ թէ Վիեննայի Մխիթարեանք:" "However, this issue is passed over in silence until now, both by the Mkhit'arists of Venice, as well as those of Vienna." Yerits'eants', *Venetiki Mkhit'areank'* [The Mkhit'arists of Venice] (Tiflis: Hovhannes Martiroseani Tbaran, 1882), 43-4. I thank Marc Mamigonian for sharing his copy of this book with me.

as a result of Trieste's active attempts to topple Venice from its pedestal of prestige as the leading port and emporium in the Eastern Mediterranean. The essay concludes by returning to the thorny question concerning the theological versus other factors behind the genesis of the Great Schism where Ormanian's arguments are placed in the context of archival and other evidence, which were not available to the former patriarch and formidable scholar.

Abbot Mkhit'ar and the Founding of the Mkhit'arist Order

Mkhit'ar Sebastats'i, as the historian Leo (Arakel Babakhanian) noted over seventy years ago, stands out among his contemporaries not only because he was truly a towering figure in early modern Armenian history but also because, unlike other prominent individuals from the period, his life has been amply and properly documented and written about.[3] There are several biographies of him written by his disciples and, moreover, we have copious epistolary documentation consisting of thousands of letters written to and by him in classical Armenian, Italian, and Latin.[4] These letters, as well as a rare contemporary chronicle of his life and the history of his congregation from its foundation in 1701 to his passing in 1749, enable us to avoid some of the pitfalls of the hagiographic elements inserted into his biography by

Figure 1. Portrait of Abbot Mkhitar Sebastats'i (1676-1749), Source: Matteos Evdokiats'i, *An Abridgment of the Life of the Lord and Most Preeminent Religious Master the Great Mkhit'ar Abbot*, Ms. 494, Mkhitarist Monastery, Vienna.

---

[3] Leo [Arakel Babakhanian], "Mkhit'areanner," *Hayots' patmut'yun*, vol. 3, 471-522 (483).

[4] Among these, the first and most important one was compiled shortly after his passing in 1749 by his trusted disciple and secretary Matteos Evdokiats'i or Matthew of Tokat, Համառօտութիւն Վարուց Տեառն Տեառն Գերապատիւ Րաբունապետին Մխիթարայ Մեծի Աբբայի [An Abridgment of the Life of the Lord and Most Preeminent Religious Master the Great Mkhit'ar Abbot], Ms. 494, Mkhit'arist manuscript library Vienna. Father Vahan Inglizian published this 80-folio manuscript in the 1949 issue of *Handes Amsorea*. It was followed by a lengthier biography by the second Abbot of the order, Step'annos Agonts' Giwver, *Patmut'iwn kenats' ew varuts' Tearn Mkhit'aray Sebastats'woy Rabunapeti ew Abbayi / hōrineal Step'annosi Giwvēr Agonts' Arhiepiskoposi ew Abbayi* [History of the life and times of the Master Mkhit'ar of Sebastea, the Master and Abbot, written by Giwvēr Agonts', Archbishop and Abbot] (Venice: San Lazzaro, 1810), and perhaps the most readable of Mkhit'ar biographies, Hovhannes Torosian's *Vark' Mkhit'aray Abbayi Sebastioy* [The life of Abbott Mkhit'ar of Sebastia] (Venice: San Lazzaro, 1901). A useful Italian-language biography by the Mkhit'arist monk Minas Nurikhan is available in English translation as *The Life and Times (1660-1750) of the Servant of God, Abbot Mechitar founder of the Armenian Mechitarists of Venice (San Lazzaro) written in Italian by Father Minas Nurikhan*, trans. Rev. John McQuillan (Venice: San Lazzaro, 1915). Finally, there is Leo's classic penetrating study "Mkhit'areanner," as well as the useful account in Ormanian, *Azgapatum*, vol. 2, 2677-2682; 2697-2698; 2703-2704; 2713-2714; 2761-2766; 2829-2834; 2947-2948; and 2969-2971.

his admirers and disciples.⁵ Taken together, they provide a fairly rich and complex portrait of a man driven by a singular passion and dedication to serve his faith and his people.

Born Manuk Petrosian on 7 February 1676 in the city of Sebastea (today's Sivas in central Anatolia/Turkey), Mkhit'ar had an early calling as a missionary and ascetic monk. After receiving a rudimentary religious education from a local Armenian parish priest, he was ordained a deacon at the age of fifteen. He soon left home as an apprentice to a traveling legate from the Catholicosate of Ejmiatsin named Ghazar Vardapet who was then passing through Mkhit'ar's hometown.⁶ His journeys and mobility in pursuit of higher education and a religious calling were not that different from those of other Armenian luminaries of his time. Like his contemporaries, Oscan Yerevants'i and Thomas Vanandets'i, Mkhit'ar was a quintessentially mobile person. He spent a good part of his life as an itinerant young monk in search of religious enlightenment and knowledge. In 1692, he accompanied Ghazar Vardapet and traveled to Tokat, then Erzerum and finally to the Holy See at Ejmiatsin, long famed as the center for Armenian higher learning.⁷ Crestfallen by corruption and widespread ignorance there, the young Mkhit'ar traveled to the monastery of the Holy Virgin, on the island of Lake Sevan. There too, disappointment greeted him. He spent the next twelve years crisscrossing the rugged terrain of Asia Minor, visiting one monastic center after another. One such place was the spiritual complex of Bassen (in the province of Erzerum, in today's eastern Turkey), where Mkhit'ar stayed for nearly a year and a half.⁸ There, two events changed the course of his life. First, he met a young Armenian missionary by the name of Poghos who had recently graduated from Rome's celebrated *Collegio Urbano*. Mkhit'ar's

---

⁵ The collection of letters in Classical Armenian is found in a rare published three-volume correspondence, *Namakani tsaṛayin Astutsoy teaṙn Mkhit'aray Abbayi eranashnorh himnadri Mkhit'arean Miabanutean* [Letter book of the Servant of God, Abbot Mkhit'ar, the blessed founder of the Mkhit'arist Congregation] (Venice: San Lazzaro, 1961). The Italian and Latin letters were published in 1980 and make up the fourth volume of his letters. See *Lettere del Servo di Dio Abate Mechitar, Fondatore della Congregazione dei Monaci Armeni Mechitaristi,* vol. IV (1705-1749) (Venice: San Lazzaro, 1980). For the chronicle see Matteos Evdokiats'i, Ժամանակագրութիւն սրբազան կարգի միաձանգն Հայոց 'ի կարգէ սրբոյն աբբայ անտոնի 'ի գերայարգոյ մխիթարայ առաքնոյ աբբայէ նորոգելոյ յորում պատմի ամենայն ինչ սրբազան կարգիս այսորիկ սկսանելով յամէ առաքնոյ կառուցման սորա եւ առ յապա, արարեալ 'ի պատուական հօրէ Մատթէոսէ աստուածաբանութեան վարդապետէ [Chronicle of the Sacred Congregation of Armenians following the rules of Saint Anthony reformed by his Eminence and first Abbot, Abbot Mkhit'ar, wherein is told all things pertaining to this Holy Order, beginning from the first year of the creation of the Order and onward, done by Esteemed Father Matteos Vardapet or doctor of theology]. Unlike the correspondence, this volume remains an unpublished manuscript in the archives on San Lazzaro. None of the studies or biographies listed in the previous footnote measures up to this eyewitness chronicle.

⁶ Matteos Evdokiats'i, Համառօտութիւն Վարուց Տեառն Տեառն Գերապատիւ Րաբունապետին Մխիթարայ Մեծի Աբբայի, 7v (In Inglizian's published version, 326). See also Leo, "Mkhit'areanner," 486.

⁷ Ibid.

⁸ Matteos Evdokiats'i, Համառօտութիւն Վարուց, folio 10r (Inglizian, 329), Leo, "Mkhit'areanner," 487-88.

imagination became illuminated by the stories he heard regarding Rome's libraries brimming with books of learning. In Bassen, Mkhit'ar also learned for the first time of the anti-Catholic persecutions launched by Avetik Evdokiats'i, the firebrand bishop of the nearby city of Erzerum, whose anti-Catholic zeal would later shift the trajectory of Mkhit'ar's mobility in the direction of Europe. These two factors deepened the young Mkhit'ar's curiosity regarding Catholicism. He returned home in 1694 but was, once again, on the road in 1695. In the spring of that year, Mkhit'ar finally met up with his destiny on the road to Aleppo, a city with nearly 8,000 Armenians, of whom only 500 were Catholics.[9] He spent three months on the outskirts of the city, boarding with the city's Jesuits, especially with the missionary Antonius Beauvollier (Antoine Beauvollier). It was under Beauvollier's influence that Mkhit'ar is said to have been persuaded that what he was seeking could only be found with Rome's help.[10]

Armed with several letters of recommendation from Beauvollier, testifying to his Catholic faith and missionary zeal, Mkhit'ar boarded a French ship headed for Rome where he wished to continue his religious education. A pestilential fever forced the traveler off the ship in Cyprus, where he endured more difficulties for several months that kept him from reaching Rome.[11] After nearly starving to death, fortune smiled upon him when a stranger paid his ship fare. He disembarked in Seleucia, on the Eastern shores of the Mediterranean and trekked, on foot, back to Aleppo. After recovering his health, he returned home to his family, and in 1696, at the age of twenty, became ordained as a celibate priest, at the church of Surb Nshan, where he first took on the name Mkhit'ar. He quickly attracted suspicion and opposition by the residents of his hometown and was forced into exile. Two years later, he was in the city that would forever transform his life.

As the imperial capital, Istanbul was a bustling and clamorous metropolis, with a large population of nearly 80,000 urban Armenians at the time.[12] Mkhit'ar had traveled

---

[9] Aleppo's Armenian population is provided by Monseigneur François Picquet, the French Consul in Aleppo and later the Bishop of Ceasaropolis. In a detailed letter to the Catholic missionary order centered in Paris and known as the *Missions Étrangères de Paris*, Picquet states that there were 8,000 Armenians in Aleppo, of whom 500 were Catholic. "Les Armeniens sont au nombre de 8 mil dont il n'y a que 500 qui sont catholiques." Monseigneur Picquet, "L'Estat de la Religion Chrestienne et Catholique dans Alep," Archives de Missions Étrangères de Paris [AMEP], vol. 352, folio 133.

[10] Nearly nothing is written about Beauvollier's influence on Mkhit'ar. His letters on Mkhit'ar's behalf are preserved and discussed in Agonts' *Patmut'iwn kenats' ew varuts'*, 78-83.

[11] Matteos Evdokiats'i, Համառօտութիւն Վարուց, folio 13r-14v (Inglizian, 333), Leo, "Mkhit'areanner," 489.

[12] We do not have the exact population figures for the Armenian community in Istanbul/Constantinople during the early modern period. My figure here is drawn from Raymond H. Kévorkian, "Le livre imprimé en milieu arménien ottoman aux XVIe-XVIIIe siècle," *Revue des Mondes Musulmans et de la Méditerranée* (September 1999), 173-85 (176). A slightly higher number of 100,000 for around the same period is provided by H. M. Ghazarian, "Merdsavor arevelk'i haykakan gaghtochakhnerě: Konstantnupolsi ew zmyurniayi gaghtochakhnerě," [The Armenian diaspora settlements of the Near East: The Diaspora settlements of Constantinople and Smyrna] in *Hay Zhoghovrdi Patmut'yun* [History of the Armenian People], vol. IV, 202.

there to meet Khachatur Erzrumets'i (known also as Khachatur Arakelian), an ardent Armenian Catholic renowned for his erudition and education as a former alumnus and missionary trained at the *Collegio Urbano*. When Mkhit'ar failed to win Khachatur over to his proposed plan of establishing a new monastic order for the education of the Armenian youth with Khachatur at its head, he once again took to the road. He traveled East to Erzerum accompanied by his two disciples. He taught at a nearby monastery called *Karmir Vank'* for a while, and, in the spring of 1700, he was back in the imperial metropolis.[13]

The twenty-five-year-old missionary found a radically transformed Ottoman capital this time. Unlike its status during the reign of the relatively peaceful Patriarch Melkisedek Suphi (1698-1699), Istanbul was now a seething cauldron of anti-Catholic persecution. Two patriarchs, Yep'rem Ghapants'i (r. 1684–1686, 1694-1698, and 1701-1702) and Avetik Evdokiats'i (1702-1703 and 1704-1706), one after the other came to the patriarchal throne with the intent of cleansing the Armenian *millet* and church of foreign accretions.[14] In part, larger forces beyond the purview of these individual actors fueled the renewal of persecution against Istanbul's 8,000-strong Armenian-Catholic community.[15] The period in which these decisive transformations in the sectarian relations between Apostolic and Catholic Armenians were occurring is called the age of "confessionalism" or "confessionalization."[16] From the shores of the Rhine and the Danube in the heart of the Habsburg empire in Europe, to the Bosporus and Zayandeh Rud in the neighboring Ottoman and Safavid Empires in West Asia, state authorities and their clerical elite were hard at work "imposing order,

---

[13] Matteos Evdokiats'i, Համառօտութիւն Վարուց, folio 18r-19v (Inglizian, 338), Leo, "Mkhit'areanner," 490, Torossian, *Vark' Mkhit'ara*, 119.

[14] For a reliable account of the Patriarchate and its holders at this time, see the republication of Hrant Asatour's, *Konstantnupolso hayerĕ ew irents' patriarknerĕ* (Istanbul: Armenian Patriarchate, 2011). See also the authoritative list of Patriarchs in the appendix of Ormanian, *Azgapatum*, vol. 3.

[15] This figure is from Charles Frazee, "The Formation of the Armenian Catholic Community in the Ottoman Empire," *The Eastern Churches Review*, 7, 2 (1975): 149-63 (153); Ibidem., *Catholics and Sultans: The Church and the Ottoman Empire, 1453-1923* (Cambridge: Cambridge University Press, 1983), 178.

[16] For a good summary of the "confessionalization thesis," see Thomas A. Brady, Jr. "Confessionalization: The Career of a Concept," in *Confessionalization in Europe, 1555-1700* (London and New York: Routledge, 2017), 1-20. For its application in West Asia, see Tijana Krstić, "Illuminated by the Light of Islam and the Glory of the Ottoman Sultanate: Self-Narratives of Conversion to Islam in the Age of Confessionalization," *Comparative Studies in Society and History* 51, 1 (2009): 35-63; Idem., *Contested Conversions to Islam: Narratives of Religious Change in the Early Modern Ottoman Empire* (Stanford: Stanford University Press, 2011); Terzioglu, "Where 'Ilm-i Hāl meets Catechism," and Guy Burak, "Faith, Law, and Empire in the Ottoman 'Age of Confessionalization' (fifteenth-seventeenth centuries): The Case of 'Renewal of Faith'," *Mediterranean Historical Review* 28, 1 (2013): 1-23. For a lucid attempt to see the Reformation in Europe in a larger context of global history, see Charles H. Parker, "The Reformation in Global Perspective," *History Compass* 12, 12 (2014): 924-934.

discipline, and religious uniformity on the population from above."¹⁷ As far as Eastern Christians, in general, and Mkhit'ar, in particular, were concerned, the most important bastion and symbol of the era of confessionalization in Europe was the establishment, in 1622, of the Sacred Congregation for the Propagation of the Faith (henceforth *De Propaganda Fide*), whose *Collegio Urbano*, founded in 1627 and its specifically Armenian college (*Collegio Armeno*) opened in 1660 in Rome, matriculated a generation of very well read but aggressively proselytizing Catholic Armenian missionaries from the East.¹⁸ The creation of new kinds of "social discipline" by state authorities and hardening of confessional distinctions between Catholics and Protestants in Western and Central Europe had their parallels and counterparts elsewhere in the early modern world. In Safavid Iran, the era of confessionalization is most famously represented by the official conversion of the Safavid state into Twelver Shi'ism, the religion of the dynasty's founder, Shah Ismail I (r. 1501-1524).¹⁹ In the Ottoman realm, confessionalization gave birth to the *Kadizadeli*s, a "group of 'puritan' preachers whose agitation and calls to religious and moral reform" helped transform the Empire into a Sunni state, all the while systematically rooting out heretical Shi'a influences or threats to Ottoman society.²⁰

---

¹⁷ Marc Forster, "With or Without Confessionalization: Varieties of German Catholicism," *Journal of Modern History* 1 (1997): 315-347 (315). The view that confessionalization was a top-down process as defined here by Foster has been modified by other scholars in recent years to include processes that are also bottom-up or driven by non-state actors. See Ute Lotz-Heumann, "The Concept of 'Confessionalization': A Historiographical Paradigm in Dispute," *Memoria y Civilización*, 4, (2001): 93-114.

¹⁸ For the history of the *Propaganda Fide*, the best study remains Josef Metzler, "Foundation of the Congregation 'de Propaganda Fide' by Gregory XV" in *Sacrae Congregationis de Propaganda Fide Memoria Rerum*, edited by Josef Metzler, vol. 1, 1622-1700 (Rome, Freiburg and Vienna: Herder, 1972), 79-111. For the Collegio Urbano, see Maksimilijan Jezernik, "Il Collegio Urbano," in *Sacrae Congregationis de Propaganda Fide Memoria Rerum*, edited by Josef Metzler, vol. 1, 1622-1700 (Rome, Freiburg and Vienna: Herder, 1972), 465-482. See also Giovanni Pizzorusso, "Una Presenza ecclesiastica cosmopolita a Roma: Gli allievi del Collegio Urbano di Propaganda Fide (1633-1703)," *Bollettino di Demografia Storica*, n. 22 (1995): 129-138, and idem., "I Sateliti di Propaganda Fide: Il Collegio Urbano e la Tipographia Poliglotta, Note di ricerca su due Istitutzioni Culturale Romane nel XVII Secolo," *Mélanges de l'Ecole Française de Rome*, 116, 2 (2004): 471-498. According to Pizzorusso ("Una Presenza," 132 and "I Satteliti," 476), there were 47 or 48 Armenian students enrolled at the Collegio during the seventeenth century alone, a figure that was disproportionately higher than that of most other ethno-religious groups.

¹⁹ See Rula Jurdi Abisaab, *Converting Persia: Religion and Power in the Safavid Empire* (London and New York: I.B. Taurus, 2004).

²⁰ Ibid., 12-3.

Given this heightened religious atmosphere across the Ottoman realms, it should not come as a surprise that the Armenian patriarchate would also get caught up in this larger wave of religious zeal among the empire's Muslim authorities.[21] In fact, the two most anti-Catholic Patriarchs of the period were directly assisted by a *Kadizadeli* and Ottoman *şeyhülislam*, Seyyid Feyzullah (1638-1703), who was not only the highest Ottoman official in charge of reforming the *ulama* of the empire but was also a personal "mentor" and "agent" of Ottoman Sultan Mustafa II.[22] Feyzullah was a key agent of confessionalization for the Empire's Armenian population. Soon after being nominated as *şeyhülislam* and in the immediate wake of the Venetian conquest of Chios (1694-1695), he helped draft the imperial edict, the *Hatt-i Şerif*, making interactions between Armenians (and Greeks) with Catholic missionaries a punishable offense.[23] The *alim* also promoted candidates to the Patriarchal throne who would pursue anti-Catholic policies at the behest of the Ottoman state. Thus with his support, Yeprem of Ghap'an ascended the patriarchal seat, for the third time, in 1701, not only driving his predecessor Melchisedek' away but also having him arrested and thrown to the galleys on grounds of secretly collaborating with Armenian Catholics and the Propaganda's missionaries operating openly in the imperial capital. Given the heightened state of affairs during the era of confessionalization where confessional and political loyalties were fused as one, Catholic Armenians (referred to as "Franks") were now widely represented by Yeprem as enemies of the Ottoman state and agents for the European powers by whose very hands Sultan Mustafa was defeated and forced to sign the humiliating Peace of Karlowitz (1699).

---

[21] For the historical context of the Armenian Catholic community in Istanbul during this time, see Cesare Santus, "La Comunità Armena di Constantinopoli all'inizio del XVIII Secolo: Scontro e tentativi di accordi Interconfessionali," *Rasegna degli Armenisti Italiani*, XVII (2016): 51-9. See also idem. *Trasgressioni necessarie. Communicatio in sacris, coesistenza e conflitti tra le comunità cristiane orientali (Levante e Impero ottomano, XVII-XVIII secolo)* (Rome: Bibliothèque des Écoles Françaises de Rome et d'Athènes, 2018, forthcoming), chap. 6, and "The Şeyhülislam, the Patriarch and the Ambassador: A Case of Entangled Confessionalization (1692-1703)," paper presented at "Entangled Confessionalizations" conference at Central European University, Budapest, June 1-3, 2018.

[22] Suraiya Faroqhi, "An Ulama Grandee and his Household," *The Journal of Ottoman Studies* (1989): 199-208 (200) and Stanford J. Shaw, *The History of the Ottoman Empire and Modern Turkey, volume 1: The Rise and Decline of the Ottoman Empire, 1280-1808* (Cambridge: Cambridge University Press, 1976), 223. My thoughts here are influenced by conversations with my graduate student Daniel Ohanian who also kindly provided me the article by Faroqhi.

[23] On the role of the *Hatt-i Şerif*, see Anna Ohanjanyan, "Gevorg Mxlayim Ōłli: An Overlooked Agent of Confessionalization," paper presented at "Entangled Confessionalizations" conference at Central European University, Budapest, June 1-3, 2018. See also Santus, "The Şeyhülislam, the Patriarch and the Ambassador." I thank both authors for permission to cite their excellent studies.

Figure 2. Citta di Modone, Coronelli Vincenzo, 1687, Source: Wikipedia Commons

Mkhit'ar was preaching at the Armenian church of Surb Gevork (Saint George) in Pera or Galata, on the European side of the capital, when Yeprem (then, only a bishop of Edirne) launched his first wave of persecutions against Catholic Armenians. Early in 1701, the bishop managed to secure the assistance of the şeyhülislam Feyzullah in obtaining control of the Patriarchate from Melkisedek and, with the help of Feyzullah and Sultan Mustafa, obtained an imperial edict for the persecution of the empire's Armenian Catholics and the arrest of their most prominent preachers in the imperial capital.[24] Along with at least two other missionaries, including Khachatur Arakelian Erzrumets'i and Sargis Tokhatets'i, Mkhit'ar was also singled out in the imperial edict.[25] He immediately took cover at the Capuchin Mission in Galata, in the same compound as the French ambassador's residence, and, therefore, came under French diplomatic protection. There "on the eighth day of September [1701], having summoned all the student monks to him, he assembled them and with them began to deliberate as to which part of the world it would be possible for them to go and establish in a safe place a habitation [for their new congregation]."[26] Mkhit'ar had originally thought of the mountains of Lebanon but soon settled on the town of Modon (Methoni) on account of its low cost of living, its proximity to Armenian-populated centers in the Ottoman Empire, as well as its relative safety from the Ottoman authorities, since it was located in the Venetian-controlled Peloponnese. Mkhit'ar took refuge there, in early 1703, shortly after a new round of persecutions was launched by Yeprem's successor, Avetik Evdokiats'i, who turned out to be more violent than the patriarch

---

[24] Matteos Evdokiats'i, Ժամանակագրութիւն սրբազան կարգի միաձանցն Հայոց [Chronicle of the Sacred Congregation of Armenians belonging to the Sacred Order of... folio 39.

[25] Ibid. For Sargis Evdokiats'i's involvement in the edict, see Grigoris Galemkearian, *Kensagrut'iwnner Erku Hay Patriark'neru ev tasn episkoposneru ev zhamanakin hay kat'oghikeank'* [Biographies of two Armenian Patriarchs and ten bishops and of Armenian Catholics of the period] (Vienna: Mkhitarist Press, 1915), 85-6.

[26] Ibid., folio 45. "Նպա յամսեանն սեպտեմբերի յալուրն ութերորդի, գամենայն աշակերտեալ միաբանսն առ իւր կոչեալ՝ անդէն ժողովէր, խորհիլ սկսօք հանդերձ, թէ յոր կողմն աշխարհիս մարթ իցէ արդեօք երթալ եւ գապահով տեղի ինչ բնակելոյ հաստատել:"

he replaced.[27] With fifteen disciples, Mkhit'ar built his congregation on Modon and even drafted a constitution for his new order in 1705. Scarcely had the building of the convent been completed, however, when war broke out. In 1715, as Ottoman forces sieged Modon, Mkhit'ar boarded a ship, once again, and traveled with his disciples to Venice armed with letters of recommendation to high officials from the Venetian governor of Modon who was a friend and admirer of Mkhit'ar. Two years later, the Venetian Senate granted Mkhit'ar and his flock the Island of San Lazzaro (a former leper colony in the Middle Ages), in the Venetian lagoon, as their permanent residence.

For the next three decades, Mkhit'ar set to work with his disciples to transform San Lazzaro into a laboratory where each of the main tenets of his congregation were put to practice. With financial patronage from Armenian merchants and other patrons, he completed the renovations on the ancient chapel on the island and built the grounds for his monks' living spaces, library, and refectory. He then continued with fulfilling his order's mission. First and foremost, he received young "novices" from different parts of Asia Minor, but predominantly from well-known families in Istanbul, and after rigorous vetting and discipline, he trained them as missionaries and scholars who were expected to travel to the East to preach the Catholic faith to Armenians. Despite the outbreak of libelous attacks on Mkhit'ar and his flock for either not being sufficiently Catholic (as was the case in 1705 and, again, in 1718, when he had to send representatives or visit Rome himself to clear himself of charges[28]) or of being apostates from the Apostolic Armenian Church, Mkhit'ar maintained an "ecumenical" view wherein he did not see any contradiction between being an Armenian, in terms of collective "national" identity, and a Catholic. Mkhit'ar best summed up this philosophy, which had guided him during his many years as a wandering preacher, in his 1733 publication of the Holy Scriptures in Venice. In the colophon of this work, he writes,

> ...though I love my nation and my labors on account of its benefit, my heart will never come loose from the orthodox faith of the Church of Rome. And conversely, though I am entirely subjected and will subject myself in faithfulness to the throne of Rome for which our father Saint Gregory the Illuminator is an example for me, my love and my striving to labor for the benefit of my nation (though it may scorn me because of such faithfulness [to the church of Rome] will never slacken. Also, if it happens that I shall be despised or condemned by everyone, or by some people, because of such things that I may have said, then I embrace them willingly. For I expect nothing from my nation and from each reader of these scriptures in return

---

[27] Ibid.

[28] These allegations are discussed in a printed pamphlet Mkhit'ar prepared for his visit to Rome in 1718. See *Eccelentissimi e Reverendissimi Signori* and *Sommario degli Attestati presentato nell'anno MDCCXVIII agli E[ccelentissi]mi, e R[everendissi]mi Prencipi i Sig[nori] Cardinali della Congregazione de Propaganda Fide Dai Monaci Armeni di S. Antonio Abate fondati in Modone, e Residenti in Venezia* (Roma, MDCCXXVIII), in *Manuscripta Italica* quart. 68, Jaggielonian University Library, Cracow, folio 3v-34r. To the best of my knowledge, scholars of the Mkhit'arists have not utilized this source before.

for my work. Rather, I desire and wish only this, that you benefit from it and shall receive the curing medicine for your souls from these Holy Scriptures..."[29]

In addition to the evangelizing and missionary elements of his congregation's objectives, Mkhit'ar also devoted his life and the energies of his disciples into rescuing and reforming the Armenian literary heritage. To this end, he transformed San Lazzaro into a "nimble and tireless workshop...into a small miniature Armenia, a homeland of books."[30] Mkhit'ar and his successors sent out missionaries to Armenian-populated regions in the East, not only to evangelize and preach, but also to "rescue" ancient Armenian manuscripts from inaccessible monastic centers and to preserve and study them back in San Lazzaro. On the basis of this project of "totalizing" archives of the past, Mkhit'ar and his followers set out to reform the modern Armenian literary canon through their publishing mission.[31]

When Mkhit'ar passed away, in 1749, his disciples appointed one of his earliest and most trusted followers, Father Eghia Matirosian (1665-1757), as an interim caretaker or *locum tenens*, until elections could be held to nominate an official successor.[32]

---

[29] Abbot Mkhit'ar, *Astuatsashunj Girk' Hnots' ew Norots' Ktakaranats'* [The Holy Scriptures, the Old and New Testaments] (Venice: Antonio Bortoli, 1733), 1279. "Չի թէպէտ սիրեմ զազգն իմ, եւ զաշխատիլն վասն օգտի նորա, բայց այնու սիրտն իմ 'ի յուղղափառէ դալանութենէ հալատոյ սրբոյ եկեղեցւոյն հռոմայ ո՛չ երբէք մեղկանայ: Եւ հակադարձաբար՝ թէպէտ ըստ ամենայնի ստորակելեմ եւ ստորաքկանեմ զիս միշտ 'ի ներքոյ հնազանդութեան զահին Հռոմայ, ըստ որում եւ Հայրն մեր սրբոյն Գրիգորիոս լուսաւորիչն ինձ օրինակ կայ, բայց այնու սերն եւ քան աշխատանացս առօգնուտն ազգին իմոյ (թէեւ զիս վասն այսպիսւոյ հնազանդութեան քամահիցէ) ո՛չ երբէք թուլանայ: Եւ եթէ նաեւ վասն այսպիսեաց իմոց ասութեանց 'ի կեանս յայս յամենից մարդկանց, եւ կամ յումանց խոտիլ եւ առ ոչինչ համարիլ ինձ առաքի կայ զայն եւս յօժարաբար [ընդ]գրկեմ. թէ իցէ այն ներկայ, եւ թէ ապագայ: Քանզի զոչինչ փոխարէն սիրոյ եւ աշխատանաց իմոց պահանքեմ յազգէն իմմէ, եւ յիւրաքանչիւրոց ընթերցողացդ զայս մատեան. այլ բաղձամ, եւ ողձակերտեմ զայս միայն, զի դուք օգտիցիք եւ զղեղս ֆրկարարս հոգւոց ձերոց յայսմանէ սրբոյ մատենէ վերընդունիցիք..."

[30] Leo, *Hayots Patmut'yun*, vol. 3, 503. Cf. Panossian, *The Armenians: From Kings and Priests to Merchants and Commissars* (New York: Columbia University Press, 2006), 103.

[31] See the excellent study by Marc Nichanian, "Enlightenment and Historical Thought," in *Diaspora and Enlightenment: The Jewish and Armenian Experience*, eds. Richard Hovannisian and David Myers (Atlanta: Scholars Press, 1999).

[32] Father Eghia's tenure as an interim successor is not discussed in any source known to me. However, unpublished letters from abroad to the island during this period address him as the caretaker.

A Coup d'État on an Island?

More than twenty years had passed during which the Abbot general ruled his cloistered family with a pacific calm when a secret conspiracy began to disturb the monastic tranquility of such a sacred enclosure and to alarm the civil and ecclesiastical authorities in Venice.[33]

Figure 3. Portrait in oil of Abbot Step'annos Melkonian (1717-1799), Source: San Lazzaro degli Armeni

On 6 April 1750, almost a whole year following Mkhit'ar's death, one of his disciples, Step'annos Melkonian, was appointed as the Congregation's second Abbot. Melkonian was away in his native Istanbul and had been immediately recalled to Venice while Mkhit'ar was still on his deathbed. The choice of appointing him was partly based on the monks' widely held belief that he was the one chosen by Mkhit'ar to succeed him. Soon after assuming his position, the thirty-three-year-old new Abbot aroused discontent among some members of his flock. At issue were his authoritarian personality, his inability to compromise with those over whom he ruled and his miserly habits as far as managing the Congregation's funds and looking after the needs of his fellow monks was concerned.[34]

Initially, Melkonian's appointment does not seem to have been for life or even for a specifically defined tenure. Even the scope of his powers was not spelled out, but we can only speculate on this, since so little is known about the island's constitution at the time. What we know with more certainty, as we shall see later, is that, sometime in the 1750s, well

---

[33] Giuseppe Cappelletti, *Storia del Cristianesimo del Prete Giuseppe Cappelletti: dall'anno 1720 a tutto il 1846, in continuazione a quella dell'Abate di Berault-Bercastel dedicata a S.M. la Regina di Sardegna* (Firenze: Tipografia di Alcide Parenti, 1847), 209. "Erano scorsi più di quattro lustri, che l'abate generale governava in pacifica calma la sua claustrale famiglia, quando una secreta cospirazione insorse a turbare la monastica tranquilità di quel sacro recinto, ed a porre in allarme la civile e l'ecclesiatica potestà di Venezia."

[34] In his authoritative four-volume *Patmut'iwn Muratean ew Haikazean Varzharanats' ew Mkhit'arean Abbayits'* (The History of the Muratean and Haikazean Colleges and of the Lives of the Abbots) published in 1866, the Mkhit'arist monk and scholar Sargis T'eodorean places blame for the schism on the island squarely on Melkonian and his personality. In volume four of this work, T'eodorean argues that on account of Melkonian's "miserly ways and insensitive habits, dissent and misfortune fell on the Congregation and caused harm to our nation." "սակա կծծի բարուց, եւ անւլողայ վարուց, վասն այսորիկ խռովութիւն եւ եղկութիւն միաբանութեան եւ վնաս ազգիս թորգոմեան:" (4: 570)

into his tenure, Melkonian seems to have felt the need for establishing a (new?) constitution and getting it ratified by the *De Propaganda Fide*. This constitution appears to have been finally drawn up in the early 1760s, following which, on 4 December 1762, Pope Clemente XIII appointed Melkonian to lifetime tenure.[35] This appointment, as well as the unexpected bequest of a substantial fortune of 100,000 piasters to the congregation in 1764 by two Catholic Armenian merchants, who passed away unexpectedly in Calcutta (India), brought simmering opposition to Melkonian to a boil.[36]

According to a nineteenth century Mkhit'arist monk Sargis T'eodorian, the priors on San Lazzaro had come to a common agreement that the money from India would serve two principal goals. First, it would be devoted to build separate living quarters on the island for the care of the older and more infirm monks who had long been suffering under penurious conditions. Second, the monks believed that the bequest would be spent towards "the establishment of a school for the education of the youth of the Armenian nation from among whom only those who had a calling for the priesthood would be recruited into the congregation, while the rest would be repatriated to their homeland to serve as enlighteners for the nation."[37] When Melkonian refused to honor these arrangements

Figure 4. Portrait of Astuatsatur Babikian (Deodato Babik) (1738-1825), Source: Wikipedia Commons

---

[35] Giuseppe Cappelletti in his *Storia del Cristianesimo* (209), published in 1847, appears to have been the first to note this. See also Hovhanness Zavrian, "Mkhit'arian Miabanut'ean Bazhanumĕ" [The Separation of the Mkhit'arist Congregation], *Hayrenik Amsagir*, No. 3 (1932), 135-136.

[36] The inheritance money was for a hundred thousand pieces of eight or silver piasters. The brothers Zaccaria and Joseph Shahriman/Sceriman, scions of one of the wealthiest families from New Julfa and ardent supporters of the congregation, bequeathed it to the order. For the history of this bequest from India and of its wealthy donors, see Sebouh David Aslanian, "Reader Response and the Circulation of Mkhit'arist Books Across the Armenian Communities of the Early Modern Indian Ocean," *Journal for the Society of Armenian Studies*, 22 (2013): 58-94 (81-9).

[37] *Patmut'iwn Muratean ew Haikazean Varzharanats'*, 4: 571. "Խորհին այնժամ միաբանքն առհասարակ եւ որոշեն հասարակաց հաւանութեամբ խնամ տանել նախ ծերոց եւ հիւանդաց, առանձինն բնակարան մի շինել, ի տածողութիւն հանգստեան ծերոց, եւ յապաքինութիւն ցաւոց հիւանդաց։ Երկրորդ՝ դպրոց մի հասարակաց հաստատել ի դաստիարակութիւն մանկանց ազգին Հայոց, եւ ընդունիլ ի ոցանէ ի կրօն միաբանութեան զնոսա միայն, որք կոչումն ունէին. իսկ մնացեալսն դարձուցանել ի հայրենիս իւրեանց, ի լուսաւորութիւն ազգին։"

and, instead, gave the fortune on an interest-bearing loan to a local Armenian merchant, Marchese Giovanni di Serpos (Յովհաննէս Սերբոսեան), the disgruntled monks took action. Thirty-nine-year-old Minas Gasparian of Artvin, who is reckoned to have been the mastermind, and the younger Astuatsatur Babikian (Deodato Babik) of New Julfa, headed the charge against the Abbot. In 1772, they called for a general assembly of the monks, formally known as a "chapter" or "capitolo" and later forced Melkonian to put his policies and his lifetime appointment to a popular vote. When the latter refused, knowing full well that a majority of ten out of nineteen monks on the island wished him to step down, the monks forcibly relieved Melkonian of his duties. Shortly afterwards, they announced to the others that Mkhit'ar's successor was no longer in power and even went so far as to place him under house arrest.[38] In a previously unconsulted first-person narrative to which we will return in the concluding section of this essay, Abbot Melkonian provides the following graphic description of what happened in a letter to the Cardinals of the *De Propaganda Fide*:

> Thus, notwithstanding [what was said above], in middle of the night, having closed the gates of the monastery and with the keys in their hands, ten of the younger monks assembled and invited all the others to join but could only persuade four others to join them. And thus holed up in the so-called chapter [capitolo] following various confused and incoherent speeches and debates among themselves, they proposed my deposition and put it to a vote. With twelve votes, so they said, they had me deposed. Then they came in full force into my room, and entering in they locked the door, taking the key from the lock. Then after giving me notice of their decree of having deposed me from my office, and of not recognizing me anymore as their superior, they ordered me to vacate my room, leaving all the insignia of my authority behind. I said to them that I have higher superiors than them on whom I depend, and without their order I do not recognize their chapter as legitimate. They replied that I absolutely had to obey and go out... so I left the chamber, but before leaving they made me leave all the keys in my office, and all alone I was left to retire to a room assigned by them where they brought me no more than just a bed. Then, they went to dinner and in the refectory they announced that I had been deposed and that in the meantime a deputy friar had been appointed until the election of new Abbot. They also told me that for one or two days neither I nor any of the monks who opposed them should leave the monastery, and go

---

[38] The first and to date most serious study of the events on the island is Carlo L. Curiel, "La Fondazione della Colonia Armena in Trieste," in *Archeografo Triestino, 1929-1930,* 339-79. This was followed by Hovhanness Zavrian, "Mkhit'arian Miabanut'ean Bazhanumě" [The Separation of the Mkhit'arist Congregation], *Hayrenik Amsagir,* no. 2 (1931), 97-109, no. 3 (1932), 133-46. Neither study seems to have attracted much attention. My account below partly relies on these excellent studies but supplements them whenever possible by directly consulting the Inquisitori di Stato folders in busta 876 and 538.

to Venice, and therefore kept the doors of the monastery closed, and the keys in their power.[39]

This brazen move opened the way for the unprecedented interference into the island's internal affairs of the secular authorities from the mainland led by the shadowy body known as the *Inquisitori di Stato*. Created in 1539 by the Council of Ten to "safeguard the secrecy of state affairs" and initially called the "Inquisitors against the disclosure of Secrets," the *Inquisitori di Stato* was a secretive body made up of a Supreme Tribunal of three magistrates and, at least, one officer in Venice known as the "footman of the Heads"[40] (*fante de cai*), whose job it was to execute the Tribunal's decisions, as well as collect and report information.[41] The Inquisitori's jurisdiction and activities covered many areas of life in Venice ranging from state security to domestic and foreign espionage. As one scholar has recently put it, "nothing and nobody escaped the ears and eyes of their spies. Their *confidenti* were ubiquitous…. and reported on anyone and anything that could pose a threat."[42] Gamblers, prostitutes, impostors, troublemakers, ambassadors, and foreign spies, all these fell under

---

[39] Archivio Storico della Congregazione di Propaganda Fide (henceforth ASPF), SC. Armeni, vol. 17, 444r. Report of Stefano di Melchiore to the College of Cardinals, 2 October 1773. "Ciò però nullaostante a mezz'ora di notte, chiuse le Porte del Monastero, ed assicuratisi delle chiavi si radunarono dieci Monaci de più Giovani, procurando che v'intervenissero pure anche gli altri tutti, ma non poterono persuaderne che soli altri quattro. E chiusi così in preteso Capitolo dopo vari confusi incoerenti discorsi e dibattimenti fra loro seguiti, proposero la deposizione dal mio carico, e posto alla ballottazione sifatto progetto si disse che con dodici voti abbiano preteso di avermi come deposto. Quindi si portarono in truppa in mia camera, ed entrati dentro chiusero la porta levando la chiave dalla serratura, e poi m'intimarono il loro decreto di avermi deposto dall'uffizio, e di non riconoscermi più per loro Superiore, e mi comandorono di uscire dalla mia camera, lasciando tutta le insegne di superiore. Risposi loro che noi abbiamo maggiori da quali dipendiamo, e senza loro ordine non conosceva legittimo il loro Capitolo. Essi mi replicarono che dovevo assolutamente ubbidire ed uscirmene. Onde risposi che essendo io solo non bastavo a contrastare a quella moltitudine (della quale, a dir vero, in tale circostanza temevo anche di qualche soverchia inconvenienza onde m'incamminai per uscire dalla Camera, ma prima di uscire mi fecero lasciare tutte le chiavi appartenenti al mio uffizio, e cosi soletto fui lasciato a ritirarmene in una camera da loro assegnatami dove non mi hanno trasportato altre che il solo letto e poi se n'andarono a cena e là in Refettorio pubblicarono la mia deposizione, avendo frattanto deputato un vicario sino all' elezione di nuovo Abbate. Mi fecero inoltre sapere, che per uno o due giorni né io né alcuno de' Monaci ad essi contrari dovessimo uscire dal Monastero, e andare a Venezia, e perciò tenevan sempre serrata le porte del Monastero, e le chiavi nel loro potere."

[40] For a brief account, see Francis Marion Crawford, *Salve Venetia: Gleanings from Venetian History* (New York: The Macmillan Company, 1905), 310-11.

[41] Ioanna Iordanou, "What News on the Rialto? The Trade of Information and Early Modern Venice's Centralized Intelligence Organization," *Intelligence and National Security*, 31, 3 (2016): 305-26 (321). See also Paolo Preto, *I servizi segreti di Venezia*, (Milano: Il Saggiatore, 1994) and the comments in the important work by Filipo De Vivo, *Information and Communication in Venice: Rethinking Early Modern Politics* (Oxford: Oxford University Press, 2007), 5, 33-6.

[42] Iordanou, "What News on the Rialto?" 321.

the watchful and seemingly omnipresent gaze of the *Inquisitori*. Even Armenian priests did not escape their attention.

Not surprisingly, soon after the removal of Abbot Melkonian, rumor of these events drifted across the lagoon and came to the attention of the *Inquisitori's* Supreme Tribunal, as well as of the Patriarch of Venice, neither of which easily brooked dissent in their realm. According to a report submitted to the Tribunal on 20 March 1773 by its most dreaded *fante de cai*, Cristofolo de Cristofoli, who represented the Inquisitors during this period, a small contingent of Venetian forces, accompanied by Cristofoli, landed on the island and immediately arrested the culprits, restoring Melkonian to power.[43] In addition to Gasparian and Babikian, Cristofoli listed the following eight other monks by name in his report: P. Gomidas Uschiudarluogh (Հայր Կոմիտաս Իսկիւտարլի), P. Luca di Simone (Հայր Ղուկաս Սիմոնեան), P. Antonio Ucicardas (Հայր Անտոն Իչքարտաշեան), P. Davide Ucicardas (Դաւիթ Իչքարտաշեան--Պոլիս 1738 -- Տրիէստ 1779), P. Pietro Mardig (Հայր Պետրոս Մարտիկեէց), P. Nicolò Pusa (Նիկողայոս Բուզայեան, 1739-1803), P. Stefano Aconz (Հայր Ստեփաննոս Կիւվեր Ագոնց, 1740-1824), and P. Paolo Meher (Պօղոս Մեհերեան, 1729-1814).[44] All ten had plotted Melkonian's downfall and were given several days to rethink their actions and repent. During that time, the island remained under total lockdown as Venetian forces cut off communications with the mainland.[45]

Christofoli paid another visit to San Lazzaro on 25 March to inform the Abbot of Patriarch Giovanni Bragadin's planned visit on 13 April. He wasted no time in carrying out the Patriarch's orders of relieving Deodato Babikian, then a lecturer of philosophy on the island, and Step'annos Agontsʻ, the superior of the novices joining the Congregation, of their duties.[46] He also learned during this visit that eight of the ten monks behind the conspiracy

---

[43] Report of Cristofolo de Cristofoli, 16 March 1773 in Inq. di Stato, b. 876. The papers in this folder are unpaginated and can be located by date.

[44] Ibid. For additional information on these monks, including the Armenian originals of their names, dates of birth, and death wherever available, I have relied on Anonym. "Mkhitʻarean amboghj harkʻ," [All the Fathers of the Mkhitʻarists] *Bazmavep* (1901): 216-26. It is interesting to note that the compiler of this list appears to have intentionally left out dates of death for those members who went over to the Trieste side and remained there.

[45] See Carlo Curiel, "La Fondazione della Colonia Armena in Trieste," *Archeografo Triestino*, 43, 1 (1929/1930): 337-79 (341). "E per tre giorni il braccio secolare custodì il convento, a fin d'impedire qualsiasi communicazione fra l'isola e la città." See also Zavrian, "Mkhitʻarian Miabanutʻean Bazhanumĕ," 136.

[46] Agontsʻ was an early coconspirator in the coup to overthrow Melkonian but decided to break ranks with Gasparian and Babikian and switched sides to being one of Melkonian's most loyal followers. He replaced Melkonian in 1800 as the congregation's third abbot and crowned his intellectual achievements by authoring a trailblazing multivolume work titled *Geography of the Four Parts of the World* (Աշխարհագրութիւն չորից մասանց աշխարհի) as well as the first published biography of Abbot Mkhitʻar, *The History of the Life and Times of the Lord Mkhitʻar of Sebastea, Religious Master and Abbot*) (Պատմութիւն Կենաց եւ Վարուց Տեառն Մխիթարայ Սեբաստացւոյ Րաբունապետի եւ Աբբայի), published in 1810 on the island.

to topple Melkonian, including Agonts', had caved in and repented. Gasparian and Babikian, on the other hand, were steadfast in their conviction. At this point, the Inquisitor's footman promptly ordered the eight remorseful coup plotters to give a full account of their actions to Patriarch Bragadin, which they agreed to do orally and in person. He also succeeded in securing the agreement of the monks to "hand over peacefully four pistols and a sword, which they told me were used to protect their monastery."[47]

Once the situation on the island had been pacified, none other than the Patriarch of Venice himself paid a ceremonial visit to San Lazzaro. Giuseppe Cappelletti, in his now-forgotten volume from 1840, *Storia del Cristianesimo*, describes the events best:

> The 13th of April finally arrived. It was the third day of the feast of Easter: in a public and solemn form, the patriarch went on his established visit, accompanied by his cortege and with the footman [i.e., Cristofoli] provided to him by the tribunal of the inquisitors of state. The Patriarch was welcomed as befitting his dignity. He celebrated the holy mass inside the church, administered the holy Eucharist to all professed clergy, novices, and laymen. After the local and royal visit, he turned to the personal visits. He began by requesting from each one [of the monks] the profession of faith and the promise of obedience to his commands, then pronounced a solemn decree by which he suspended from any sacred ministry all and each of the rebellious and seditious monks who had dared to attempt so much against their rightful superior, and threatened them with even more serious canonical penalties, even by means of the secular authorities if in the future they refused to render to the same Abbot their due obedience. Following such a decree, the guilty ones threw themselves at the feet of the patriarch, displaying signs of repentance, begging forgiveness for their enormous error, and uttering words of respect and submission to the Abbot. With this they were freed from being suspended from clerical services. However, the threat of suspension was not fully lifted; on the contrary the culprits were more strictly inculcated with

---

[47] Report of Cristofolo de Cristofoli, 27 March, 1773 in Inq. di Stato, b. 876. "E così da me essendo stato fatto, ed a tutti intimato, dopo averli insinuati alla doverosa, e rispettosa obbedienza verso il loro Capo, e per aderire à pubblici comandi e per oggetto della comune lor pace mi furono consegnate da essi quattro Pistole ed una sabla che mi dissero servire per custodia de lor Monastero, e che perciò due erano già del convento medesimo e due gli erano state recate in dono pel medessimo fine dal Signor Serpos, ed in appresso un Archibugio, il quale pur dissero essergli stato dato da tenere in salvo da un forestiero. Lo che tutto rassegno di E.E. V.V. in conformità dei loro venerati comandi. 1773, 27 Marzo."
"And, thus, once I reminded all of them of their dutiful and respectful obedience towards their leader, I ordered them to adhere to the public commands and for the sake of the public good, peacefully they handed over four pistols and a sword which they told me were used to protect their monastery. For this reason, two pistols were already in the same convent and two were given for the same purpose by Signore Serpos. In addition, there was also an arquebus, which they also said, was given for safekeeping by a foreigner. All of this I report to you in conformity to your venerable orders. 1773, 27 March."

the obligation to obey their superiors and warned of the prohibition of further seditious gatherings.[48]

The fate of the two recalcitrant ringleaders, who presumably had demurred from offering the Patriarch his wished-for "promise of obedience to his commands," was sealed more than a full month after the Patriarch's visit. Once again, it was left to Cristofoli to impose stern discipline. On the evening of 16 May, a patriarchal secretary (*cursore*) and the Inquisitor's footman arrived at the monastery and, in the presence of seniors of the congregation, had the two above-mentioned monks summoned. Cristofoli then ordered the two to return and place in the hands of the Abbot the letters of recommendation (*lettere credenziali*) they had violently taken from him as well as "all the seditious correspondence with the individuals of the Monastery and also with some other persons from the outside."[49] According to Cappelletti, Father Minas obeyed this command, but Babik refused to do so "pretending not to understand the orders with vain excuses."[50] He was finally subdued, and it was discovered that two Armenians outside of the congregation had also played a role in

---

[48] Cappelletti in his *Storia del Cristianesimo,* 211-12. "Giunse alla fine il dì 13 Aprile: era la terza festa di Pasqua. In forma pubblica e solenne si portò il patriarca alla visita decretata; lo accompagnava la sua corte; il tribunale dell'inquisitori di stato lo aveva sussidiato altresì del suo fante. Fu accolto come alla sua dignità convenivasi; entro in chiesa, celebrò il santo sacrifizio, amministrò la santa Eucaristia a tutti i cherici professi, ai novizi, ai laici. Fatta la visita locale e reale, passò alla personale. Cominciò coll'esigere da ciascheduno la professione di fede e la promessa di obbedienza ai suoi comandi. Quindi pronunziò solenne decreto con cui sospese da qualunque sacro ministero tutti e ciascheduno dei monaci ribelli e sediziosi, che avevano osato attentare cotanto contro il loro legittimo superiore, e li minacciò di ancor più gravi pene canoniche, anche per mezzo del braccio secolare ove in avvenire avessero ricusato di prestare al medesimo abate la dovuta obbedienza. In seguito di siffatto decreto vennero i colpevoli ai piedi del patriarca, monstrando segni di pentimento, implorando dell' enorme lor fallo, e proferendo parole di rispetto e di sommesione all'abate. Con ciò ottenero d'essere sciolti dalla sospensione; ma non fu tolto, anzi fu loro viepiù strettamente inculcato l'obbligo dell'obbedienza al loro superiore e la proibizione di ulteriori combriccole sediziose."

[49] Cappelletti, *Storia del Cristianesimo,* 212. "Quindi il fante del tribunale intimò loro il supremo ordine di restituire e depositare nelle mani dell'abate stesso le lettere credenziali, che a lui violentemente avevano carpito, nonché tutti li scritti di sediziosa corrispondenza cogl' individui del monastero e con qualunque altra persona del di fuori."

[50] Ibid. "Ubbidì a questo comando il padre Minas, ma il Babik se ne ricusava or fingendo di non intendere ora cercando sottrarsi con vane scuse. Vi fu alla fine costretto, e si scoprì, che alla loro scandalosa congiura avevano preso parte anche due armeni estranei alla congregazione, il prete Michele di Murat e il secolare conte Zaccaria Sceriman."

their scandalous conspiracy: the priest Michele di Murat (see below) and the secular count Zaccaria Sceriman.[51]

The patriarchal representative then gave the Abbot two decrees from the patriarch, ordering the summary expulsion of the two subversive monks. Gasparian and Babikian were therein "declared reckless, seditious, suspended from church services, demoted from the rank of monks, and condemned to dress in the habit of secular Armenian priests."[52] The respective decrees of suspension (*sospensione a Divinis*) were even given to each of them, and the patriarchal punishment was carried out. Afterwards, gondolas arranged to transport the two convicts were instructed to wait while they went to their rooms to gather every scrap of paper containing writings that were deposited in a sealed chest (*in un baule suggellato*) and handed over to the Abbot.[53] These papers, it seems, were subsequently placed in the State Archives of Venice where they are still preserved.[54] Once the paperwork of the seditious monks was confiscated, Cristofoli himself accompanied the perpetrators to the threshold of exile. The first to be banished was Deodato Babikian, who was taken by the

---

[51] Count Zaccaria Sceriman (also spelled Seriman) was a well-known literary figure in Venice and a member of the Julfan Armenian family of diamond merchants who had settled down in Venice in 1698 but had branches residing in Iran, the Russian Empire, as well as India. It was two brothers of this family's branch in Calcutta who had left the large bequest to the Mkhit'arists in 1764 (see above). For the family's history, including its family tree, see Sebouh David Aslanian, *From the Indian Ocean to the Mediterranean: The Global Trade Networks of Armenian Merchants from New Julfa* (Berkeley: University of California Press, 2011), 149-58 and Sebouh D. Aslanian and Houri Berberian, "The Sceriman/Shahrimanian family of Julfa," *Encyclopedia Iranica* (2009). For Zaccaria's life, but without any mention of his involvement in the 1773 events in San Lazzaro, see D. Maxwell White, *Zaccaria Seriman: The Viaggi di Enrico Wanton, a Contribution to the Study of the Enlightenment in Italy* (Manchester: Manchester University Press, 1961).

[52] Ibid. "Ivi ciascuno di essi nel decreto, che lo riguardava, era dichiarato temerario, sedizioso, sospeso dal sacro ministero, degradato dalla condizione di monaco, e condannato a vestire l'abito di prete armeno secolare."

[53] Report of Cristfolo Cristofali, May 16, 1773, Inq. di Stato, b. 876. "...indi essendo loro stata intimata dal cursor Patriarcale la sospensione a Divinis, e fatti spogliare delle Divise da Monaci, furono condotti uno per volta in separata Gondola e quivi fermaronsi sino a che ritornato alle loro rispettive stanze raccolsi ogni carta continente scrittura che portava in un baulle suggellato consegnai all'Abate."

[54] I discovered these confiscated materials at the Archivio di Stato di Venezia in 2005. They are stored in two separate boxes in ASV, Miscellanea di Atti Diversi, Manoscritti, filza No. 106-I and 106-II. They contain about twenty of Gasparian's private letters with his father in Khotorjur in the vilayet of Erzerum, dating back to 1753, as well as manuscript drafts of an Armenian Dictionary and a theological treatise. Yet they do not seem to include any correspondence with the two mainland co-conspirators mentioned by Cappelletti. The letters of Ter Hovsep' Gasparian, Minas Gasparian's father, to his son, spanning from 1753 to 1769, are in the first box in an envelope titled "Le Lettere dirette da Don Giuseppe Gasparian al P. Minas (Mechitarista?)"

footman straight to a boat that was to transfer him to Trieste.[55] Father Minas Gasparian was taken to Trento, and both were ordered by supreme decree to perpetual exile from the city of Venice and from all the domains of the Serene Republic.

Almost two weeks later, Cristofoli returned to the island to finalize the Patriarch's decision of forgiving the eight repentant monks and allowing them back into the order. His report of 2 June 1773 describes his visit to San Lazzaro on Saturday, 29 May.

> I the undersigned footman of the Supreme Tribunal report that on Saturday 29 of the last May I went to the island of the San Lazzaro degli Armeni where I summoned the Abbot and made known to him the order that I had from your most excellencies, namely of immediately freeing the eight monks well-known to him. Once everybody [meaning the eight monks] came before him and myself, I explained to them the will of your excellencies of graciously granting them the restoration of their liberty on the assertions of their father Abbot regarding the repentance of their errors, hoping that they would continuously confirm their obedience and due respect towards their Superior. At which expression, they immediately threw themselves before him and confirmed their profound submission together with an extraordinary happiness.[56]

---

[55] Report of Cristfolo Cristofali, May 16, 1773, Inq. di Stato, b. 876. "e ritornato poscia alle Gondole notai prima sopra questa, in cui si trovava il P. Diodato Babigh, e mi feci condurre alla Barca stabilita per Trieste intimando al di lei Padrone chiamato Tommaso Lizza di trasportar detto Padre a Trieste, e intimando pure al Padre istesso di non più trasferirvi negli stati del Dominio veneto per comando Supremo di V.V. E.E."

[56] Report of Cristofolo de Cristofoli, June 2, 1773 in Inq. di Stato, b. 876. "Riferisco io sottoscritto fante di codesto Supremo Tribunale come Sabato 29 Maggio decorso mi trasferii all' Isola di S. Lazaro degli Armeni, ove fatto chiamare l'Abate feci a lui noto il Comando che avevo di V.V. E.E. cioè di rimettere nella prima Libertà gli otto Religiosi ad esso ben noti; per lo che intervenuti tutti alla presenza di lui e di me significai loro la volontà del'E.E. V.V. di concedergli benignamente ripristinazione in libertà sulle asserzioni del loro Padre Abate riguardo al ravvedimento degli incorsi errori, sperando che sempre più confirmarebbono in essi l'obbedienza, ed il rispetto dovuto verso il lor Superiore. Alle quali espressioni si prostrarono immantimenti innanzi a lui, e diedero contrassegni sul fatto di profonda rassegnazione insieme e estraordinaria allegrezza."

The very next day, the footman took care of one more unresolved business, namely the matter of the Armenian priest on the mainland, Don. Michiel Murat (Հայր Միքայէլ Մուրատեան). It had become clear from the correspondence seized from the "seditious" monks the week before that Murat was complicit in the conspiracy to unseat Melkonian from the start.[57] On account of this, the Tribunal had resolved to exile the monk for life from "Venice and the domains of the Republic" and had given him "between three and eight days" to leave Venetian territory, risking "the penalty of death if he ever returned."[58] This set the stage for the Trieste chapter of the Mkhit'arists' history.

Trieste and Giacomo Casanova

Situated astride the intersection of Latin, Germanic, and Slavic worlds and ideally located to take command of the maritime trade of the Eastern Mediterranean, the region of Trieste (formerly the Venetian territory of Tergestum) had come under the rule of the Habsburg dynasty in 1382.[59] For much of its early history, the city was an unimportant fishing village on the northern armpit of the Adriatic Sea.[60] In the early eighteenth century, it experienced a rapid transformation. Under the Holy Roman Emperor Charles VI of Austria, Trieste, along with Fiume (Rijeka), was transformed in 1719 into a "free port," at a time when Venice's grip as the regional maritime economic power in the Eastern Mediterranean was loosening. Eager to supplant the former Queen of the Adriatic, Charles VI presided over large-scale infrastructural reforms that improved the transportation and maritime networks of Trieste and made the city into an ideal hub and early modern duty-free port, with incentives

---

[57] Not much is known about Michiel Murat. According to Sahak Djemjemian, Murat was born in Istanbul in 1729 and ordained a priest on San Lazzaro in 1752. Along with a fellow monk, Gabriel Bedrosian, he was expelled from the island two years after being ordained, following orders from the *De Propaganda Fide*, the Venetian authorities, and, of course, Abbot Melkonian himself. Sahak Djemjemian, *Hovhannes Patkerahan: Namakani* (1695-1758) (Venice: San Lazzaro, 1988), 264 n. 245. It appears that there were several priests expelled from San Lazzaro during the first decade of Melkonian's tenure.

[58] Ibid. "Domenica poi giorno seguente trovato personalmente il religioso D. Michiel Murat di Nazione Armeno gli intimai che per comando di V.V. E.E. debba partir da Venezia fra il termine di giorni tre e di giorni otto dallo Stato della Repubblica sotto pena della vita se più tornarà, e ciò tutto eseguito in obbedienza à comandi di V.V. E.E.1773 2 Giugno."

"The following day, on Sunday, I personally found the priest Don. Michiel Murat of the Armenian nation and ordered him that by the command of your excellencies, he had to leave Venice and the domains of the Republic between three and eight days under penalty of death if he ever returned and therefore everything was carried out in obedience to the commands of your most excellencies." 1773, June 2."

[59] Lois C. Dubin, *The Port Jews of Habsburg Trieste* (Stanford: Stanford University Press, 1999), 1 and 2

[60] Ibid., 2.

meant to lure commercially savvy "minority" communities.⁶¹ Trieste also became one of the nodes for the Ostend Company, the Habsburg counterpart to the other European East India Companies, and was made into the "seat of an emergent Austrian navy."⁶² In addition to making the city into a kind of "tax-free zone," the emperor extended religious toleration to non-Catholic communities such as Sephardic Jews, Orthodox Christians (including Greeks and Serbs) as well as Lutheran Germans whose numbers had risen to 46 in 1756.⁶³ The results were impressive. From a small town of barely 5,000 citizens at the start of the eighteenth century, Trieste's population grew by a remarkable fivefold by century's end.⁶⁴

---

⁶¹ See Daniele Andreozzi, "Innovations, Growth and Mobility in the Secondary Sector of Trieste in the Eighteenth Century," in *Innovation and Creativity in Late Medieval and Early Modern European Cities*, eds. Karel Davids and Bert De Munck (Surey: Ashgate, 1988), 337-354, and Michal Wanner, "The Establishment of the General Company in Ostend in the Context of the Habsburg Maritime Plans, 1714-1723," *Prague Papers on the History of International Relations* (2007): 33-63

⁶² Wanner, "The Establishment of the General Company in Ostend," 35.

⁶³ Giuseppe Occioni-Bonaffons, "I vostrii bisnonni o Trieste nel secolo XVIII," [Your Forbears or Trieste in the XVIII Century] *Archeografo Triestino*, N.S. 18 (1892): 436.

⁶⁴ Ibid., 15, and Aleksej Kalc, "Immigration Policy in Eighteenth-Century Trieste," in *Gated Communities?: Regulating Migration in Early Modern Cities* (London and New York: Routledge, 2012), 117.

The desire to oust Venice from its place as a famed commercial emporium knew few bounds for the Habsburgs. They granted privilege upon privilege to entice to their tax-free haven trading communities with important services to offer. In 1742, Greek Orthodox merchants first began to frequent the city, and six years later they were sizeable enough to have their own church. That same year, Trieste saw the arrival of Orthodox Slavs from Bosnia and Dalmatia.[65] As Lois C. Dubin's impressive study of the Habsburg port's Jewish community has illustrated, by the mid-eighteenth century, several hundred Jews (mostly Sephardim and by far the city's largest and most notable minority) had moved to the Habsburg port.[66] By contrast, until the late 1760s, not a single Armenian was registered as domiciled there.[67]

In 1769, the Habsburg court in Vienna undertook efforts to remedy this situation. A Catholic Armenian priest named Giovanni Ariman[68] (Յովհաննէս Արիման or

---

[65] Oreste Cuppo, "I Padri Mechitaristi in Trieste," *La Porta Orientale: Rivista Mensile di Studi sulla Guerra e di Problemi Giuliani e Dalmati* (1931): 132-72 (132-33).

[66] For the changing figures of the city's Jewish community, see Dubin, 21.

[67] There is, however, an isolated tombstone of an unknown Catholic Armenian priest, Martino Carabeth, buried in one of the city's cathedrals in 1756. See Anna Krekic and Michela Messina, *Armeni a Trieste tra Settecento e Novecento: L'impronta di Una Nazione* (Trieste: Civico Museo del Castello di Sann Giusto, 2008), 11.

[68] Giovanni Ariman is an obscure figure in Armenian history. He was born in Kesaria (now Kayseri in Eastern Turkey) and studied under the Armenian Archbishop, Sargis Saraffian, the primate of Ankara. He seems to have been in search of gainful employment as priest at the Armenian church of the Holy Spirit (*Surb Hogi*) in Amsterdam in the late 1730s and early 1740s, because, in 1743, a Catholic Armenian confidant of Abbot Mkhit'ar in Rome, Hovannes Patkerahan, informed the Abbot that Ariman had just arrived in Rome from Amsterdam with the intent of giving an official profession of the Catholic faith and eventually travelling to Venice to meet Mkhit'ar and become a member of his order. Hovannes then recalls how Ariman created discord in the community there and eventually left. He appears not to have been accepted into the order by Mkhit'ar on account of already being twenty-eight years old. In a letter from Vienna, dated 13 December 1766, we learn that he was among the Armenians in Varadin (Transylvania) for a while before heading to Lvov and finally returning to Vienna. See Ghevond Tahyean, *Mayr Divan Mkhit'arean Venetko i Surb Ghazar, 1707-1773: Tsakmanē ukhtis minjev i bazhanumn Triesta* [Grand Archive of the Mkhit'arists of Venice in San Lazzaro, 1707-1773: From the Origin of this Order until the separation of Trieste] (Venice: San Lazzaro, 1930), 238-239. Two or so years later, he was permanently set up in Trieste as the city's parish priest for the few local Armenian Catholic residents. For his arrival in Rome in 1743, see Djemjemian, *Hovhannes Patkerahan: Namakani*, 110-111. "Նաեւ եկեալ իսկ է տեղս Մրսրտամու ումն քահանայ. Յովհաննէս Վարդապետն կոչեցեալ. ինքն կեսարու գեղեն է եւ աշակերտ Սերքիս եպիսկոպոսին, որ էր ենկուր առաքնորդ: Այս անձ տեղեակ իսկ է ուղղափառութեան բայց ոչ ողղաբարապէս [sic] վարկեալ է. Վասն որոյ յատկապէս Ռոմ է` առ ի դալանութիւն տալ եւ տնօրինումն առնելոյ: փոքր ինչ ինձ յայտնեցաց իւր դիտմունքն, զի (յ)ետ Զատկին կամի վենետիկ գալ` առ ի կռօնաւոր լինիլ ընդ Հայրութեան Ձերում (որպէս յուսադրեալ է զինքն Տէր Ստեփաննն): Բայց ես այս բանիս դժվարութիւնն ինչ պարզաբանեցի իւրեան։ Տարօք 28 ամաց է:".

Արիմանեան) communicated the first signs of this new policy of establishing a colony of Armenian entrepreneurs in the Habsburg free port in his missive of September 12 of that year to Abbot Melkonian in Venice. "In order to turn this place more populous," writes Ariman in his letter, "the Empress has established Trieste as a free city. To this end, [the court] considered that if they were to maintain in Trieste an Armenian priest, it would facilitate the settling down of Armenians here, so that those Armenians who wished to reside in this place would not be deterred from doing so on account of the absence of a priest."[69]

The residence of the above letter-writer in Trieste, in the fall of 1769, is considered to be the start of the Habsburg port's tiny Armenian community.[70] Starting from that year, a few "co-nationals" did indeed respond to Ariman's call by applying for naturalization papers, and several even set up residence in the city. The most notable of these was Giorgio Sarraf or Գէորգ Սարաֆեան (originally from Edessa or Urfa), who, beginning in 1770, was the director of Trieste's *Compagnia d'Egitto* and acted as the city's official interpreter

---

[69] This letter is reproduced in part in a selection of archival documents from the Mkhit'arist archives. See Ghevond Tahyean, *Mayr Divan Mkhit'arean Venetko i Surb Ghazar, 1707-1773*, 261. "Կայսերուհին ազատ կացոյց զքաղաքն Դրիեստին վասն բազմամարդ առնելոյ զնա, այսպէս առ այս վախճան խորհեցան հասատատել ի Դրիեստէ եւ զհայ քահանայ մի, զի գուցէ եթէ իցեն յազգէն հայոց, որք կամիցին գալ, եւ բնակիչ լինիլ ի Դրիեստէ, վասն պակասութեան քահանային մի արգելցին ի գալոյ, վասն այսորիկ աՀա առաքեաց զիս ի Դրիեստէ տալով ինձ զերկու Հարիւր ֆիորինս եկամուտ յամին բացի ողորմութենէ պատարագին մինչեւ զերիս ամս. զի տեսանելով, թէ յաճախիցին Հայք ի գալ ի Դրիեստէ ի մէջ այսչափի ժամանակիս, ունիցիմ միշտ զայն երկու Հարիւր, կամ աւելի եւս եկամուտն: Արդ տեսանելով Գերյարգելութեան ձերոյ զՀաստատիլն աստ Հայ մի մշտնջենական քահանայի միոյ մերոյ գալստեամբ ազգայնոց ի Դրիեստէ, Հաճեցի ուրեմն ադաչեմ ըստ նախանձու բարեաց հլրոց օգնել առ այս, եւ գրել յարեւելս առ միաբանս Գերյարգելութեան ձերոյ, զի յորդորեսցէն զմերայինս առ ի գալ, եւ Հանգչիլ ի Դրիեստէ ցուցանելով նոցա զազատութիւնն, զպատշաճաւորութիւնն վաճառականութեան եւ զդիւրագին ապրուստն տեղւոյն, ունելով եւ զմխիթարութեան իւրեանց Հոգելոր որշափի ես ապրիցեմ. եւ յետ մեռանելոյն իմոյ կարեն ապահով լինիլ մերայինքն ունիլ ի քաղաքայ իւրեանց ումն ի միաբանից Գերյարգելութեան ձերոյ՝ ծանուցանելով իմ՝ ի ժամանակի իւրում դրան կայսերական զարդիւնս վանականաց Գերյարգելութեան ձերոյ ի յոլել զհայս աստ՝ եւ զարժանաւորութիւն լինելոյ ի ճմերայնց աստ՝ ի քաղանայ Հայոց... : 1769 ի սեպտեմբերի 12 ի Դրիեստէ Նուաստ ծառայ Յոհաննէս Արիման քահանայ անարժան."

I thank Merujan Karapetyan for this reference.

[70] See Kalc, "Immigration Policy in Eighteenth-Century Trieste," 121-122, and Liana de Antonellis Martini, *Portofranco e Comunità Etnico-Religiose nella Trieste Settecentesca* (Milan: Dott. A. Giuffrè editore, 1968), 142-143.

Figure 5. Portrait of Joannes Ter Raphael Babikian, Source: *William Robertson, Vipasanut'iwn Ameriko (History of America)* vol. 1 (Trieste, 1784)

for Oriental languages.[71] Hovhannes Mkrtich Sarum, known as Giovanni Battista di Sarum, father of Samuel Mkrtich Mooratian the wealthy benefactor of Venice's *Collegio Moorat Raphael*-fame, also applied for naturalization papers in 1774 for himself and his two sons, Samuel and Carapiet, requesting from the city's Chamber of Commerce (the *Intendenza Commerciale*) the privilege of "settling down in this free port with the intention of undertaking my usual career of commerce."[72] Mr. Sarum appears to have resided in Trieste for only one year. He moved to Venice the following year and, in short order, relocated with his sons to India, where many years later, as we shall see, one of his sons, Samuel, became a great benefactor for the Mkhit'arist Congregation.[73] In that same year, Trieste welcomed as its latest naturalized Armenian Joannes Ter Rafael Babikian (Giovanni di Raffael), the brother of Deodato Babikian, one of two "seditious" monks exiled from Venice and later

---

[71] Not much is known about Sarraf or Saraff. For scattered references, see Curiel, "La Fondazione della Colonia Armena in Trieste," 346, 348 and especially 370, fn. 33. Sarraf passed on suddenly at age 63, in 1782, and was buried at the Mkhit'arist Church in Trieste. A manuscript containing the births, baptisms, and deaths in Trieste between 1775 and 1809 and preserved in the Mkhit'arist monastery in Vienna (Ms. 454, folios 35 r-36v) has the following entry regarding this merchant: Յամի տեառն 1782. Դեկտեմբեր 29. յալուր կիրակէի 'ի 6 ժամու առալոտու փոխեցալ առ տեառն յանկարծական մահուամբ գէորգ սառաֆն առանց խոստովանութեամբ եւ հաղորդութեամբ, եւ յերկույին ձնացեալ թափորան 'ի գիւղն նորա յորում մեռեալ էր բերեաք զնա փառալորապէս յեկեղեցին մեր, եւ թաղեալ զնա ասացեալ զճարկաւոր աղօթս եւ սաղմոս, 'ի վախին՞ զկնի երեկոյեան ժամու 'ի 6 թաղեցաք զնա 'ի եկեղեցւոյ մերոյ 'ի գերեզմանի հասարակաց: Թիլ ամաց ցուցա երելի 63 ամաց:

Ես Հ. Ղուկաս Սիմոնեան կրօնաւոր Մխիթարեան հաստատեմ եւ վկայեմ:

"In the year 1782 on December 29, on Sunday at 6 o'clock in the morning, passed onto the Lord, in a sudden death and without a confession or sacraments, Mister Giorgio Saraf. We went to his town where he died in the evening and brought his body to rest at our church, and having read his rites and said ritual prayers and Psalms, we buried him solemnly after 6 o'clock in the evening in our church in a common grave. He appeared to be about 63 years of age."
I Father Ghukas Simonian affirm and bear witness to this."

[72] C.R.S. Intendenza Commerciale per il Litorale in Trieste, b. 594, folio 269 and Tullia Catalan, "Cenni sulla presenza armena a Trieste tra fine Settecento e primo Ottocento" in *Storia economica e sociale di Trieste*, vol. 1 a cura di Roberto Finze e Giovanni Panjek, 604. "...stabilito in questo porto franco con animo morandi per intraprendere il solito mio carriere di Commercio."

[73] Sahak. V. Ter Movsesian, *Murat-Rap'ayelyan varzharanneru bererarnerě yev irents ktakneṙ: zoyg krt'akan hastatut'yeants' hariwrameakin art'iw, 1836-1936* (Venice, San Lazzaro, 1937), 36-7.

settled in Trieste. Mr. Ter Rafael, or Padre Rafael as he was known in London and Calcutta, was a Julfan merchant of considerable renown and wealth. He had resided in Bengal and Surat in the East Indies for twenty-four years and was embroiled in a high-stakes and historic trial in London that nearly brought down the English East India Company.[74] After winning his trial in 1771, he had chosen to settle down with his family in Venice, where his younger brother Deodato was an up-and-coming Mkhit'arist priest. His decision, in October of 1773, to abandon Venice and seek residency in Trieste should not come as a surprise, given the expulsion of his brother in May of that year. In the eyes of many, it was also a harbinger of things to come, at least, as far as Trieste's relationship with Venice was concerned. Slowly, a nucleus of a tiny mercantile community was forming in the Adriatic port, thus making the Habsburg authorities ever more eager to lure to their growing city Armenian merchants from neighboring ports of Venice, as well as Istanbul and Izmir. The timing of Babikian and Gasparian's exile from Venice and their later arrival in Trieste aligned with this important turning point in Trieste's relationship with both Venice and Armenian merchants.

Babikian arrived in Trieste on May 19, a few days after embarking on a ship departing Venice for Trieste. His fellow monk Gasparian, who presumably had found out about his co-conspirator's destination and travelled there directly from Trento, soon, on June 9, 1773, joined him. Only two weeks after their arrival, the defrocked monks received support from Trieste's two leading Armenians, the priest Ariman and the influential and wealthy businessman, Saraff. Through them and accompanied by supporting affidavits testifying to their good credentials as *bona fide* Catholic priests sent to them by Mkhit'arist missionaries

---

[74] Shortly after meeting him for the first time on 16 September 1777, governor Karl Graf Zinzendorf wrote in his diary that "In the afternoon, Ricci brought over the Armenian Babich, who resided for 24 years in the East Indies and has won a lawsuit in London against the East India Company, and who has an English air about him. His trial is written about in the political history of Bengal." "Après-midi, Ricci m'amena l'Arménien Babich qui a vécu 24 ans aux Indes Orientales, qui a gagné à Londres un procès contre la Compagnie des Indes, qui a l'air tout anglois. De son procès il est parlé dans l'histoire civile du Bengale." Karl Graf Zinzendorf, *Europäische Aufklärung Zwischen Wien und Triest. Die Tagebücher des Gouverneurs Karl Graf Zinzendorf, 1776-1782*, vol. 2 (Vienna: Böhlau Verlag, 2009), 1777, 39. The trial in question occurred before Parliament in London and involved four Armenian business partners, or *gomasthas*, of William Bolts, the Dutch merchant and investor in the English East India Company. They were summarily arrested and tortured in Faizabad (Bengal), where they were trading, by the English governor of Bengal, Harry Verelst, in 1767. Two of the four Armenians made history by traveling to London in 1769, at the behest and with the coordination of their senior partner, Bolts, and filed a highly sensational lawsuit against the Company and its governor, which they won in 1777. The case was heard before parliamentary sessions and became celebrated through many books on the Company's reputation for corruption. It acted as a pretext for the limitations imposed on the Company by Parliamentary intervention. For a reliable account of the case, see Willem G. J. Kuiters, "Law and Empire: The Armenians Contra Verelst, 1769-1777," *The Journal of Imperial and Commonwealth History*, 28: 2 (200): 1-22, and ibidem, *The British In Bengal, 1756-1773: A Society in transition seen trough the biography of a Rebel: William Bolts (1739-1808)*. The book, "political history of Bengal," mentioned by Zinzendorf is most likely William Bolts's famous work published only two years earlier under the title *Considerations on India Affairs, Part II: Containing a complete vindication of the author from the malicious and groundless charges of Mr. Verelst*.

in Elizavetpol in Transylvania with whom they had maintained a traffic of letters, the two presented their case to the port city's Chamber of Commerce:

> We the undersigned testify as the truth to whom it may concern that Fathers Minas Casparian and Deodato Babich are well known to us as being people of exemplary conduct and also of hailing from a wealthy family of merchants. We also testify that we know a congregation of pious monks of the Armenian nation have gathered and built a convent in Venice where they attract novices of Armenians from the East and where they have built a school and where they print various books in the Armenian language, and that from this convent missionaries travel to the East to convert the heretical Armenian nation to the Roman Catholic religion. Both in Venice and the East, these monks lived off the gifts and donation left to them by many Armenian nationals in such a manner that the monastery became celebrated (insigne) only thanks to the wealth bequeathed to them from the East. And given that humble congregations could be established wherever, they [the two monks] have made their desire to establish a similar congregation in this free port known to the religious authorities. And not lacking in neither knowledge of the Armenian and Latin languages nor the riches of their wealthy relatives, as we have noted, they have requested this very attestation and we, for the simple truth without any ulterior motives have made them the present certificate written in our own hand. Trieste 22 July 1773.

Giovanni Ariman
Armenian priest
Giorgio Saraff
Imperial Interpreter[75]

---

[75] The Austrian State Archives (*Österreichisches Staatsarchiv*) AT-OeStA/FHKA NHK Kommerz Lit Akten 724 Armenier und armenische Mechitaristen in Triest: Religionsexerzitium und Geistliche (25/1), 1749-1792, folio 28. "Noi infrascritti attestano per verità à chi aspetta qualmente conosciamo benissimo li R.R. PPP Don Minas Caspariens, e Don Diodato Babick, essere d'ottima condotta e vita Esemplare, ed anche d'esser di Famiglia ricca de Negozianti, che conosciamo altresì, che una radunanza, o sia Congregazione Semplice di pie Persone radunati in Venezia della nazione Armena hanno eretto un Convento, ritirando in esso Novizi Armeni Orientali, in dove hanno eretto le Scuole, stamparono vari libri in lingua Armena, e che da questo convento andarono nelle Missioni nell'Oriente, convertendo la nazione Armena Eretica, alla Religione Cattolica romana, vivendo quanto in Venezia che in Oriente dalle lasciate fatte da molti nazionali Armeni in maniera che divenuto un Monastero Insigne, solo con beni Orientali; E siccome le semplice Congregazioni possono ovunque sorreggersi, così ci hanno communicato li Prefatti Religiosi l'animo loro l'erreggere in questo Porto Franco una simil Congregazione, non mancando ai medesimi ne scienza della lingua Armena, e Latina, neppur mezzi de'Parenti facoltosi, come a noi sono noti perciò ci hanno pregato, di rilasciarli presente Attestato, e noi per pura verità senza fine alcuno li abbiamo il presente Attestato coscrivendolo col proprio pugno. Trieste li 22 Luglio 1773
Giovanni Ariman
curato Armeno
Giorgio Saraff
Interprete Cesare"

It is unclear whether the authorities in Venice anticipated that with an official *sospensione a Divinis* from the Patriarch, the monks would be deprived of their only source of livelihood, namely preaching or engaging in religious work of any sort. What is certain, however, is that, much to their chagrin, news arrived from the Venetian consul in Trieste, Marco Monti, that Babikian and Gasparian had not only joined forces in Trieste and were thriving but were slowly beginning to attract to Trieste the *crème de la crème* of Venice's Armenians. Moreover, they were also secretly communicating with sympathizers back at the convent in San Lazzaro as well as with the latter's monastic branch in Transylvania. That this was in fact happening is conveyed in an alarming letter to the Tribunal of the Inquisition from none other than Abbot Melkonian, sometime probably in the summer or fall of 1774.

> As a matter of fact, the above-noted two expellees have not only tried to separate from this place some Armenian families who have settled in this city along with their funds and trades that are advantageous to the state [of Venice], but they are also attempting to seduce even the resident monks of Venice to detach themselves from the body of this community and to unite with them to reinforce their faction. They have carried this out by means of a suspected secret correspondence through which they have successfully lured Padre Davide Ucicardas [Հայր Դաւիթ Իւշտաշեան]. The latter, being sent by me to Constantinople after a convenient request he made to me, arbitrarily went against the duties of his obedience and disembarked in Trieste to join with them.[76]

The priest mentioned here, Davide Ucicardas, appears to have been an ingenious inventor and is credited with casting printing matrices and punches for Armenian letters with which the Trieste faction was able to set up its own press in 1775. He was able to leave San Lazzaro after being covertly contacted by a certain Hakopjan di Hermet, a Julfan physician in Venice who was then the trusted medical doctor for the monks and, as such, had relatively unrestricted access to the island, despite a communication embargo in place after the coup.[77] Hermet's insistence to the Abbot that Ucicardas needed to change climate

---

[76] Undated letter of Stefano Melchiore (Stepʻannos Melkonian) to the Tribunal of the Inquisition (most likely written in the later part of 1774). ASV Inq. di Stato, b. 876. "Di fatto i suddetti due espulsi oltre distaccamento che tentano di fare di alcune Famiglie Armene stabilite in questa Capitale co i loro fondi e commerci vantaggiosi allo stato; cercano sedurre anche i Monaci qui dimoranti e distaccandoli dal corpo di questa Comunità, unirli seco loro a rinforzo della loro fazione: Siccome è riuscito ad essi per via di secreto carteggio (di cui si ha probabile sospetto) adescar il P. Davide Ucicardàs, il quale essendo stato da me per le sue importune spedito per Costantinopoli andò arbitrariamente, e contro doveri della sua obbedienza a sbarcar in Trieste, e collegarsi con loro."

[77] The role of Hakob (Hakobjan) Hermet in facilitating the defection of Ucicardas is discussed in Curiel, "La Fondazione," 347. For Hakob Hermet and his family, see Vahram Torgomian, "Kensagrakan: Hakob kam Hakobjan bzhizhk Hermetian" [Biography: Hakob or Haskobjan Physician Hermetean] *Handes Amsorea*, 5 (May 1904): 133-39. See also ibidem, "Kensagrakan: Petros bzhizhk Hermetian" [Biography: Petros Physician Hermetean] *Handes Amsorea*, 7 (July 1904): 213-17. See also Anna Krekic and Michela Messina, *Armeni a Trieste tra Settecento e Novecento: L'impronta di Una Nazione* (Trieste: Civico Museo del Castello di Sann Giusto, 2008), 26-27.

without delay in order to be cured from his (feigned) illness persuaded Melkonian to allow the priest to board a ship to his native Istanbul.

The surreptitious traffic of secret letters continued undetected after Davide Ucicardas's near-literal jumping ship, in October of 1773. Two more defectors from San Lazzaro succeeded to join their brethren in the Habsburg port the following year. The catalyst for their departure was, once again, a member of the same family of Julfan physicians and, in this case, the Basra- or Julfa-born Petros Hermetian or Pietro Hermet who, like his father Hakob, had free access to the island as a medical doctor.[78] During the summer and fall of 1773, both Melkonian, as well as the authorities on the mainland, appear to have been in a state of crisis. Venice's attraction to Armenian merchants as a cosmopolitan seaport was hanging in the balance. One spy report after another, drifting in from Trieste, underlined the danger posed by the two exiled Armenian monks. An anonymous report, sent from Trieste on 25 November of 1773, warned the Tribunal of the "great destruction" ("la gran rovina") that Babikian and Gasparian were causing to Venice, as well as San Lazzaro, noting how Giorgio Sarraf was assisting them in every possible way. The author of the report emphasized how he was moved by his "zeal for the public good and for justice to notify [the following goings on] to your Excellencies so that you may put a stop to such harm that by the day grows bigger and threatens the welfare of the above-mentioned convent and also of the Armenian nation whose life depends on the subsistence of the Armenian fathers in that city."[79] Among the dangers listed in his report are the slow but worrisome departure of well-to-do Armenian families (such as that of Joannes Ter Rafael) from Venice and other ports in the Levant to the Habsburg port and the reluctance of Armenian merchants to bequeath funds to Melkonian and his flock.[80] The outcome of the reports prompted the Tribunal to appeal to Venice's consul, Monti, for assistance. In the spring of 1774, the latter hatched an unlikely plot to bring the work of these obstreperous preachers to an end. He suggested

---

[78] Melkonian explains this, in his letter to the Inquisitors. See undated letter of Stefano Melchiore, ASV Inq. di Stato, b. 876. "E al presente due altri Monaci di questo Monastero il P. Antonio Ucicardàs, e 'l P. Niccolò Pusa, ravveduti già de loro passati trascorsi, ed acquietati, mossi dalla fama della nuova impresa de sudetti espulsi, e forse anche eccitati nel segreto abboccamento tenuto con un tal Pietro di Giona, de cui con ragionevol fondamento si può sospettare d'esser stato da colà specialmente incaricato per tal oggetto o con lettere o con verbal commissione si cambiarono in un tratto, e sollecitarono instantemente a disfarsi da questa Communità, e andarsene come si presume a Trieste: benché dal canto mio non lasci intanto di adoperare con essi tutti i Religiosi e caritatevoli mezzi per distorli dalla loro precipitosa risoluzione, e forse ravveduti si rimetteranno a miglior via."

[79] Anonymous report sent from Trieste on 25 November of 1773, ASV Inq. di Stato, b. 876. "...son in obbligo per il zelo, di ben Publico, e di giustizia, avvisar la vostra Eccellenza per mettere argine a tanto male giornalmente dilatato alla rovina del sopraddetto convento, ed anche della nazione Armena la di cui vita dipende quasi dalla sussistenza de suddetti Padri Armeni di cotesta città."

[80] Ibid. "E di più moltissime famiglie Armene vengono a Venezia per l'amor de Padri Armeni di S. Lazaro, quando dunque sentiranno il guaio della loro discordia cagionata da questi due Padri, aborriranno da venire, e piuttosto quelle che sono venute si tentavano tornar a dietro."

Figure 6. Giacomo Casanova (1725-1798)
Source: Wikipedia Commons

to the *Inquisitori* that they hire as a spy a Venetian adventurer who had recently arrived in Trieste, after spending twenty years doing the "Grand Tour" across European capitals while in exile from his native Venice. This adventurer was none other than the internationally renowned womanizer, adventurer, and man of letters, Giacomo Casanova. According to one nineteenth-century commentator, it "seemed to the Consul that Casanova who had free access to the house of Saraf could take back the Armenians to San Lazzaro, unburdening the Republic of great anxiety."[81]

Since his infamous escape from the *Piombi* prison in 1756, Casanova had, on several occasions, tried to ingratiate himself with the Venetian authorities by offering to put his services as a "secret agent" and mole at their disposal in exchange for a general pardon for himself, allowing him to return to his city of birth. For instance, while residing in London, in November of 1763, he volunteered to work as an informant for the Venetian ambassador there but with no concrete results.[82] When the opportunity arose to work for the *Inquisitori* in Trieste, in the fall of 1774, he readily accepted. The result was that he wrote a steady stream of reports to his spymasters that have survived in the Venetian state archives and provide a rare window into the activities of the renegade monks in Trieste. Casanova also devoted the last chapter of his twelve-volume French autobiography, *L'histoire de ma vie* (*The History of My Life*), to his time as an infiltrator, paying visits and befriending Trieste's newly arrived monks, who, unbeknownst to authorities in Venice, were in the process of purchasing a large villa and convent and, with the technical support of Davide Ucicardas, were preparing to establish their own printing press—something that even their mother convent would not have till at least 1789.[83] In

---

[81] Occioni-Bonaffons, "I Vostrii Bisnonni o Trieste nel Secolo XVIII," 443. "Parve al Console che il Casanova, il quale aveva libero accesso nella casa del Saraf, potesse assumersi di ricondurre gli armeni a S. Lazaro, togliendo alla Republica un motivo di grande inquietudine."

[82] Iordanou, "What News on the Rialto?," 321. See also Paolo Preto, "Giacomo Casanova and the Venetian Inquisitors: A Domestic Espionage System at Work in Eighteenth-Century Europe," in *The Dangerous Trade: Spies, Spymasters, and the Making of Europe*, ed. Daniel Szech (Edinburgh: The University of Edinburgh Press, 2010), 139-156 (148).

[83] The best account of Casanova's role as a spy in Trieste remains that of Carlo Curiel, "La Fondazione della Colonia Armena in Trieste," *Archeografo Triestino*, 43, 1 (1929/1930): 337-79. Preto, "Giacomo Casanova and the Venetian Inquisitors," is poorly informed on the Trieste affair.

chapter ten of volume twelve of his memoirs, Casanova informs his readers that the Tribunal of the *Inquisitori* had given Consul Monti a hundred silver ducats to be offered to him "to encourage me...that I could hope for everything from the clemency of the Tribunal if I could resolve the great problem of the Armenians, of which the Consul could give me the details."[84] He then provides the relevant background, adding that having failed in their objective of making the expelled monks "return to Venice" by "direct means, that is, through the action of their Abbot," the State Inquisitors, according to Casanova, had taken recourse to "secret means to raise obstacles to them in Trieste which would discourage them from remaining there."[85] This is how the last major adventure of his eventful life, his final "secret mission," was hatched. Casanova warmed up to the monks "by striking up an acquaintance with them on the pretext of going to see their Armenian types, which they had already had cast, and a stock of precious stones and minerals which had come to them from Constantinople." He succeeded in winning over their trust in only a week. It did not take too long for him, however, to realize that this mission was futile. The monks spurned his friendly recommendations of begging the Abbot's forgiveness and returning to the bosom of their order in Venice. On the contrary, they became more adamant and explained to him that they might consider reconciliation if the Abbot were to "recover the four hundred thousand ducati which he had entrusted to the Marchese Serpos at four per cent interest."[86] In addition, the monks also demanded that Abbot Melkonian institute certain reforms to the governance of the monastery, a topic on which Casanova does not elaborate. In any event, this last part proved impossible to resolve, thus leaving Casanova's mission a dismal failure.

Perhaps the most useful takeaway from Casanova's account of the genesis of the Trieste order is the lucidity with which he addresses the real stakes of the intrigue. The Habsburgs not only acceded to Gasparian's and Babikian's request for asylum at once "but also granted them privileges," Casanova writes. The Venetian philanderer and *bon vivant* makes it abundantly clear his *History of My Life* that, as far as the Habsburgs were concerned, it was "a matter of ousting Venice from her place in this branch of commerce [the printing of Armenian books] and giving it to Austria."[87] The documentation on these monks preserved in the State Archives in Vienna makes it evident that the principal reason as to why these two expellees had little trouble in acquiring a captive audience with the Habsburg court in Vienna and its representatives in Trieste, such as the city's governor (the famed count Karl Zinzendorf), was the allure of commerce, not only of printing but also of long-distance trade with the East in which Armenians in the Mediterranean basin had long established a reputation as being necessary go-betweens. In a petition addressed to the Chamber of Commerce, in October of 1773, the monks detail the reasons for their expulsion from the Monastery of San Lazzaro, to which we will turn later, and then make a solid case for why

---

[84] Giacomo Chevalier de Seingalt Casanova, *The History of My Life*, translated by Willard R. Trask (Baltimore and London: Johns Hopkins University Press, 1997), 199.

[85] Ibid., 200.

[86] Ibid., 201.

[87] Ibid., 200.

their settlement in the Habsburg port city would be beneficial to the Austrian authorities:

> Two [priests] in particular moved to this free port with the idea of taking with them their rich families, partly residing in Venice and partly still in the East. In this manner, [they believed] it would be easy for those who are here to recall their wealthy families to settle down in this free port and that this would be more so the case given that the Armenian nation loves having its priests at hand [wherever it is settled]. In a manner similar to how the congregation was created in Venice, the monks thought it would be possible to do the same here.[88]

This petition had its desired effect. In 1775, Empress Maria Theresa issued a special edict granting singular privileges to Babikian, Gasparian, and two others, who had since jumped ship from Venice, recognizing them all as Austrian subjects and allowing them to build a new branch of the Mkhit'arist Congregation.[89] The most important privileges in this 53-article edict, issued on 30 May 1775, concern the right of the fledgling congregation to run its own press (articles 24 to 30), to enjoy a thirty-year monopoly of printing and selling books in the Armenian language, to freely practice their Uniate Catholic faith, and to operate a school for the children of their nation.[90] Interestingly reminiscent of the East India Company's 1688 "Treaty with the Armenian Nation," the edict also granted Armenian merchants the privilege of being full Austrian subjects, of owning and leasing ships in the maritime trade of Trieste

---

[88] AT-OeStA/FHKA NHK Kommerz Lit Akten 724 Armenier und armenische Mechitaristen in Triest: Religionsexerzitium und Geistliche (25/1), 1749-1792, folio 31. This petition like many in this file was written in Italian and reads as follows: "Due Particolarmente si sono trasferiti in questo Porto Franco, con l'Idea di condurre secco la loro ricca Famiglia parte Dimorante in Venezia, e parte ancora in Oriente a guisa tale, che loro qui stando facile sarà a loro di richiammare Case buone di stabilirsi in questo Porto Franco molto più che la nazione Armena ama avere li suoi Sacerdoti onde simil Congregazione che fu fatta in Venetia potrà farsi ancora qui aspettando pro interim alcuni loro Parenti di Venezia, e da altrove per consultare e dar principio alla loro pia idea."

[89] The edict exists in at least two copies. See AT-OeStA/FHKA NHK Kommerz Lit Akten 724 Armenier und armenische Mechitaristen in Triest: Religionsexerzitium und Geistliche (25/1), 1749-1792, folios 85-95, and Archivio di Stato di Trieste (henceforth AST), C.R. Governo in Trieste (1776-1809), 1068, unpaginated folder of documents. The other monks mentioned in the edict, Padre Zaccaria d'Alexan, P. Lucas Simon, P. David Ucikardas, and Gomidas Garabiet. The edict also mentions the participation two prominent Armenian families who had moved to Trieste at this time ("e delle Famiglie Babik e Saraff"). To date, the only study of this edict is also the first brief investigation and Armenian translation of the edict's clauses in H. S. G. "Mariam T'ereziyayi T'riesti Hayots' tuats artonagirĕ," *Handes Amsorea* 6 (May 1889): 92-94. The full edict was published in its original Italian in 1861, in "1775, Statuti e Regolamento della Nazione Armena di Trieste," [Statutes and Regulations for the Armenian Nation of Trieste] in *Raccolta delle Leggi Ordinanze e Regolamenti Speciali per Trieste publicata per Ordine della Presidenza del Consiglio* (Trieste: Tipografia del Lloyd Austriaco, 1861), 12-14.

[90] Ibid., 91v, r. "Confermiamo alla Congregazione Mechitarista il Privilegio esclusivo della stampa in Trieste de' libri in idioma Armeno per trent' anni, riservando la conferma alla sovrana autorità."

and of flying the imperial Austrian colors (articles 40 to 48).⁹¹ The edict appears to have been issued with the anticipation that the ranks of Armenian merchants in Trieste would quickly swell and "great multitudes" of Armenians would settle down in the Adriatic port. In this connection, article 50 makes the provision to allow the Armenian nation in Trieste to run its affairs in accordance to its own constitution, to elect a "governor or secular head," and to have representation on the city's Chamber of Commerce.⁹²

The upshot of these privileges was to be a boon for the new branch of the congregation and a dismal letdown to the Hapsburgs. Instead of seeing multitudes arriving to its shores, or of attracting many Armenian families from neighboring Venice, only a total of thirty to forty Armenians set up shop in the Adriatic port.⁹³ The long-sought-for printing press began operating the same year as Maria Theresa's edict. It churned out considerable number of books, many addressed to readers in the Ottoman capital and written in Armeno-Turkish, the vernacular language most accessible to the greatest number of Armenian readers in Asia Minor.⁹⁴ Clearly the monks were expecting to turn a quick profit through their publishing enterprise, and that is why perhaps they began to cater also to the new secular readership across the port cities of the early modern Armenian diaspora, offering them books on creative and novel topics and unfamiliar histories. As early as 1783, they also started to

---

⁹¹ For the 1688 Treaty with the English East India Company, see the classic study by Ronald Ferrier, "The Agreement of the East India Company with the Armenian Nation, 22nd June 1688." *Revue des Études Arméniennes* n.s. 7 (1970): 427–443; ibidem, "The Armenians and the East India Company in Persia in the Seventeenth and Eighteenth Century." *Economic History Review* 2nd ser., 26 (1973): 38–62; and Sebouh David Aslanian, "Julfan Merchants and European East India Companies: Overland Trade, Protection Costs, and the Limits of Collective Self-Representation in Early Modern Safavid Iran," in *Mapping Safavid Iran*, edited by Nobuaki Kondo (Tokyo: Tokyo University Press, 2016).

⁹² AT-OeStA/FHKA NHK Kommerz Lit Akten 724, folio 94v. Article 50 states: "Dopocché si saranno moltiplicati li stabilimenti in Trieste delle famiglie armene al segno, che costituischino un Corpo Nazionale, li conferiamo ora per allora la facoltà di eleggere sotto il Presidio di un Consigliere della Nostra Intendenza Commerciale un Governatore, o Capo secolare, e due Assistenti o Deputati..."
"If the numbers of the settled Armenian families multiplies from a few to the point where they constitute a corporate nation, we will then confer upon them the right to elect, under the authority of a counselor of our Chamber of Commerce, a governor or secular head and two assistants or Deputies..."

⁹³ Catalan in her archivally grounded study provides a figure of between 26 and 39 Armenian residents in Trieste around the time of Maria Theresa's edict "Nel 1776 risultavano immigrati in città 39 armeni, e l'anno seguente se ne contavano 26." Catalan, "Cenni sulla presenza armena a Trieste," 606. In contrast, and without citing any sources, H.S.G. in "T'riesti hay gakht'akanut'iwně," *Handes Amsorea* (February, 1889): 22-3 (22), claims an amplified figure of 25 to 30 Armenian families or 100 to 150 individuals by the end of the century.

⁹⁴ On the little-studied printing endeavors of this branch of the Mkhit'arists, see Sahak Djemjemian, "T'riesti Mkhit'arean tparaně" [The Mkhit'arist printing house of Trieste] *Handes Amsorea* (1981): 75-110; on Armeno-Turkish publications and the Trieste Armenian printing press, see Sebouh David Aslanian, "Prepared in the Language of the Hagarites: Abbot Mkhit'ar's 1727 Armeno-Turkish Grammar for Vernacular Western Armenian," *Journal for the Society of Armenian Studies* (2017): 54-86.

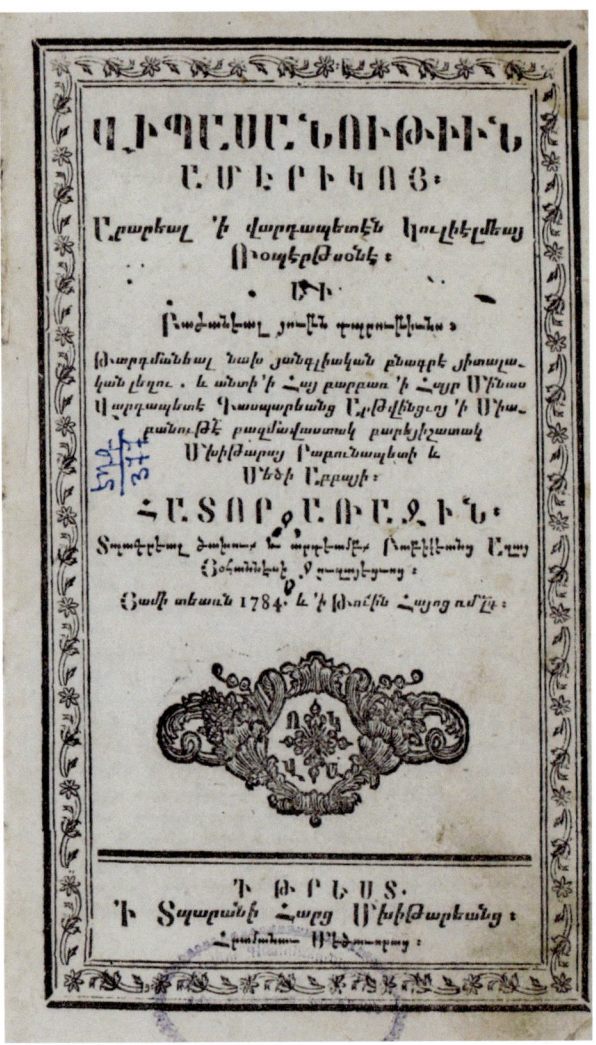

Figure 7. Frontispiece, William Robertson, *Vipasanut'iwn Ameriko (History of America) vol. 1* (Trieste, 1784)

accept for publication works of translation from European languages, such as the two-volume *Vipasanut'iwn Ameriko* (1784), the *History of America*, recently published in London by the Scottish Enlightenment figure William Robertson and specially commissioned for publication by none other than Joannes Ter Rafael Babikian whose life we have touched on above. The Trieste fathers also published the translations from various languages, including English, French, and Italian by the Julfan merchant Marcara Shahrimanian. These bold new works of literature written or translated by the new merchant, "nouveau literate," class counted among their titles Pétis de la Crois's *Histoire du Grand Jenghizchan* of 1710, published as *The History of Genghis Khan the Great, the first Emperor of the Former Moghuls and Tatars, Comprising of Four Letters* (Trieste, 1788), and Louis le Comte's (1655–1728) popular *Lettres sur les Cérémonies de la Chine* published in Trieste in 1783.[95]

---

[95] These merchant-writers, who were on the whole not formally trained like the church-educated erudite scribes or *vardapets* (*archimandrites*) who dominated the profession of writing in Armenian society, were the Armenian counterparts to the eighteenth century writers in the world of Islam that Dana Sajdi describes as belonging to "*nouveau* literacy." The latter were authors of "unusual backgrounds" who during the eighteenth century had entered "into the space that had historically been arrogated to the '*ulama*,' 'the people who know.'" Dana Sajdi, *The Barber of Damascus: Nouveau Literacy in the Eighteenth-Century Ottoman Levant* (Stanford: Stanford University Press, 2015), 6. The title of Louis le Comte's work is Թուրք գչնաց or Թուրք պատուական Հօր լուդովիկոսի Քօմթեանց շուրջ գորպիսուլ[թեամ]բ երկրին Սինեցւոց, որ է Չին, կամ Չինումաչին

However, neither these publications nor their hopes of earning income by running a grammar school for the youth of Armenian merchants in Istanbul appears to have paid off. Facing an impending financial crisis, Babikian and Gasparian had no choice but to fall back on support from Armenian merchants. In 1785, they sent a delegation of two monks, Fathers Nikoghayos Puzayan and Poghos Mēhērian (both in the group of ten who joined Gasparian and Babikian's plot to overthrow Melkonian in 1773 but had only later managed to join them in Trieste) to India in search of a generous endowment.[96] This mission is mostly important in history not because of the money it raised for the fledgling congregation but because it led indirectly to the establishment of the most important institution associated with the name of the Mkhit'arist Congregation, namely the *Collegio Armeno Moorat-Raphael* established by money given by Eduard Raphael Gharameants' and later Samuel Mkrtich Mooratian, the first of whom was contacted by Puzayan and Mēhērian in 1785 and left a large endowment to the Trieste monks as anticipated. This money was predicated on the classical Armenian translation of Charles Rollin's *Histoire romaine depuis la fondation de Rome jusqu'à la bataille d'Actium, c'est à dire jusqu'à la fin de la République* (Roman history from the foundation of Rome to the battle of Actium, that is, until the end of the Republic), which Eduard Raphael had entrusted to the monks in Trieste in exchange for a promise of endowment for a college established and managed by the Trieste monks. The monks, in particular Minas Gasparian who had undertaken the task of translation, accepted the initial money but reneged on their promise. Eventually, a considerable fortune was bequeathed to Puzayan and Mherian by Eduard Raphael in his will of 1791, but the money ultimately found its way back to Melkonian and the Mkhit'arists in San Lazzaro, when Mēhērian and Puzayan decided to jump ship again, this time back to their mother convent in Venice.[97]

In the midst of these financial difficulties, Babikian and Gasparian also sent one of their representatives (Father Anania Jambazian, 1732-1803) to Istanbul to raise money there for the struggling order. Jambazian took out huge loans in the Ottoman capital by

---

[96] Their journey to India is told in a recently published travelogue written by Mēhērian. See *Patmut'iwn Varuts' Hayr Poghos Vardapeti Mēhērian Sharagreal yiwrmē 1811, Venetik, i Vans Srboyn Ghazaru* [History of the Life of Vardapet Poghos Mēhērian, written by himself in 1811, Venice, at the Monastery of Saint Lazarus], edited by Gevorg Ter Vardanian, part 3, *Bazmavep*, 2007, 8-130.

[97] For Raphael's will and its provisions on establishing a college in Europe, see "Last will and testament of Edward Raphael, Esq. of Madras, 1792," in British Library, India Office Records (IOR) "Madras Wills," L/AG/34/29, #125, folio 49. The most authoritative account of the history of the Raphael and Muratian wills and how the money ended up going from the Trieste faction back to Venice is Sargis Teodorian's magisterial and authoritative, *Patmut'iwn Muratian ew Haykazian Varzharanats' ew Mkhit'arian Abbayits'* (Paris, 1866) 4 vols. See also the more accessible but derivative account in Barsegh V. Sargisian, *Erkhariwrameay krt'akan gortsuneut'iwn Venetkoy Mkhit'arean Miabanut'ean* [Bicentenary of the Educational Enterprise of the Mkhit'arist Congregation of Venice] (Venice: San Lazzaro, 1936), especially chapters 8 and 10. I have also covered the story in detail in my unpublished essay, "Silver, Missionaries, and Print: A Global Microhistory of Early Modern Networks of Circulation and the Armenian Translation of Charles Rollin's *History of Rome*."

issuing promissory notes payable in Trieste to certain families from the capital's wealthy Armenian community. The order used the money to purchase real estate in the city including grounds for their school. However, it defaulted in its payments and, facing a growing number of creditors, even tried briefly, through Jambazian himself, to enter talks with Melkonian's successor as Abbot, the Transylvanian Abbot Step'annos Agonts', to negotiate for reunification with the mother convent on condition that the Venice branch step in and pay back the Trieste faction's debts in Constantinople. This overture, which must have represented a low point in Babikian's life, was spurned. After Jambazian passed away in Constantinople, in 1803, things worsened for Babikian and his dwindling congregation because many of the Trieste faction's creditors, who had till then remained silent or patient, requested a return on their loans.[98] The lenders sued the monks in the courts but were blocked by the authorities in Vienna, who shielded their Mkhit'arist subjects. Things further deteriorated when Trieste, once again, came under Napoleon's jurisdiction in 1809 and was ceded to his empire following the Treaty of Schönbrunn signed in that same year. Unable to be protected by their Habsburg patrons and earlier turned away by the mother convent and with their real estate confiscated and auctioned off for a fraction of its real value by the city authorities, Babikian and a new collaborator named Father Aristakes Azarian (Gasparian had presumably passed on by then) were forced to desert the Adriatic port to seek asylum in imperial Vienna.[99] On 5 December of 1810, Babikian (by then an old man) and Azarian received from Emperor Francis I of Austria an abandoned Capuchin monastery in Vienna, at the same spot where the order exists today.[100] Babikian, till then the *de facto* leader of the new congregation in Trieste, was only formally nominated Abbot of the new order in 1803. He thus became the first Abbot of the Vienna branch, even while the dynamism of leading the Vienna branch from the brink of bankruptcy to financial security and even the lap of luxury was now in the hands of the younger Azarian. Under the latter's guidance, the new order was soon graced with imperial support for the establishment of a printing press. Once again, special privileges from the Habsburg court (including a thirty-year monopoly over the printing of religious prayer books or missals in Latin and German used throughout the empire) paved way for the order to thrive and blossom.[101]

Not long after the creation of the Vienna branch of the order, it had become apparent to all but a few that the Great Schism of 1773 had led the two branches of the Mkhit'arist Congregation not only to diverge geographically and culturally (with Venice being firmly in the Italian sphere of cultural influence while the Viennese one moved in the orbit of German culture) but also theologically. Following Babikian's death in 1826, and with

---

[98] My information in this section draws heavily from Zavrian, "Mkhit'arian Miabanut'ean Bazhanumĕ," 72-78 and T'eodorean, *Patmut'iwn Muratean ew Haikazean Varzharanats'* vol. 4, 575-81.

[99] Curiel, "Fondazione," 365-366, Nerses Akinian, "Aknark mĕ viennayi Mkhit'arean miabanut'ean hariwrameay gortsuneut'ean vray, 1811-1911," *Handes Amsorea* (1911), 7 and Zavrian, "Mkhit'arian Miabanut'ean Bazhanumĕ," 75.

[100] Akinian, "Aknark mĕ viennayi Mkhit'arean miabanut'ean," 7 and Curiel, "Fondazione," 366.

[101] Zavrian, "Mkhit'arian Miabanut'ean Bazhanumĕ," 76.

Abbot Azarian at the helm, the Viennese Mkhit'arists soon began to accuse their Venetian counterparts of being "schismatics" and not genuine Catholics. The acrimony between the two factions had reached such new heights that some at the time believed Azarian to have been personally behind the shutting down, under the orders of the censors in Rome, of the printing press in San Lazzaro for three years in 1814, right around the time that Azarian had established his profitable press in Vienna.[102] This sectarian infighting between the two factions, culminating in outright religious feuds in Constantinople, Smyrna, and elsewhere in the Ottoman Empire during the first half of the nineteenth century, compels us to turn, in our concluding section, to a central question with which we began this essay, namely what role did theological and other factors play in giving rise to the Great Schism of San Lazzaro?

Theology or Despotism? The Genesis of the Great Schism Reconsidered

The complex question of what drove the Great Schism of 1773, thus bifurcating the Mkhit'arist Congregation, has been approached from several different and, at times, irreconcilable perspectives. Generally speaking, two sets of views, or "schools" of thought, have emerged over the last century. Each is represented by a formidable giant of twentieth-century Armenology, namely Patriarch Maghakia Ormanian, on the one hand, and the Mkhit'arist scholar from the Vienna branch, Father Nerses Akinian, on the other. Before offering a critical assessment of their respective positions, it is necessary first to provide a brief stocktaking. In his 1911 survey essay on the history of the Vienna branch of the congregation, Akinian addresses the schism almost in passing and describes the events and their causes as he sees them.

> Shortly after his election, Melkonian undertook to alter the canons established by the blessed founder [i.e., Mkhit'ar]; however, with these new changes of his, he gave rise to discontent among the most senior members of the congregation. For seven years, he pursued in vain to get the constitution he established ratified in Rome until the discontent that grew day by day compelled the members of the congregation to call for a general assembly on 25 May 1772, where the Abbot [Melkonian] was himself invited (*while still not established as such by Rome*) and where the restoration and preservation [*verahastatut'iwn ew pahpanut'iwn*] of the old constitution was demanded. However, since it was not possible to arrive at an agreement, Melkonian was brought down, and it was decided to hold elections for a new Abbot. But before a new session could be convened, Melkonian appealed to external intervention and with the involvement of the archbishop of the city, he attempted to compel the subordination and loyalty to his will of his subjects. Two archimandrites, fathers Astuats'atur [Deodato] Babikian and Minas Gasparian, who stood firm on the resolutions of the assembly, were escorted by a contingent of troops and taken out of the country and managed to come ashore in Trieste…

---

[102] Ibid., 77.

where the bishop of Trieste and the municipality graced them with all the [relevant] facilities.[103]

Akinian's argument above that the constitution and its breach by Melkonian were at the heart of the troubles that befell San Lazzaro is sound enough to compel us to ask if, indeed, the island actually had a constitution to begin with, or whether, as Ormanian has insinuated, it was ruled by Mkhit'ar largely "on the basis of personal authority."[104] We know from one of Mkhit'ar's biographers, Hovhannes Torossian, that as early as 1705 Mkhit'ar had drafted a set of rules or a constitution for his newfound order which was based on the

---

[103] Akinian, "Aknark mě viennayi Mkhit'arean miabanut'ean hariwrameay gortsuneut'ean vray, 1811-1911," *Handes Amsorea* (1911): 4. Emphasis added. "Մելքոնեան իր ընդրությենէն քիչ ետքը ձեռնամուխ կ՛ըլլայ երանաշնորհ Հիմնադրին սահմանած կարգերը փոփոխել. սակայն իւր այս նորածեւությիւններով յառաջ կը բերէ երիցագոյն միաձայն մէկ դժգոհություն։ Եօթը տարի կը հետապնդէ ի զուր իւր կազմած Սահմանադրությիւնը Հռոմ հաստատել տալ, մինչեւ որ օր աւուր աճող դժգոհությիւնը կը ստիպէ միաբանները՝ ընդհանուր ժողով մը գումարել։ 1772 Մայիս 25ին կը բացուի ժողովը. ուր կը հրաւիրուի նաեւ Սրբահայրը (տակաւին չհաստատուած Հռոմէն) եւ կը խնդրուի վերահաստատութիւն եւ պահպանութիւն հին սահմանադրութեան. բայց որովհետեւ կարելի չ՛ըլլար համաձայնութիւն մը գոյացնել, Մելքոնեանը վար կ՛առնուի եւ կ՛որոշուի նոր Սրբայի ընտրութեան ձեռնարկել. բայց դեռ նոր նիստը չբացուած՝ Մելքոնեան կը դիմէ արտաքին ուժի եւ քաղաքին Արքեպիսկոպոսի միջամտութեան. կը փորձէ զինուորական զօրութեամբ ստիպել իւր հպատակներն ի հնազանդութիւն իւր կամաց:

Երկու միաբաններ--Հ. Աստուածատուր Վ. Բաբիկեան եւ Հ. Մինաս Վ. Գասպարեան--որոնք ժողովոյն որոշմանց վրայ հաստատուն կը մնան, զինուորական ուղեկից գնդով արտասահման կը հանուին: Ատոնք կը յաջողին Տրիեստ ցամաք ելել:

Սակայն ի Սուրբ Ղազար այսու խաղաղությիւնը չի վերահաստատուիր. Երկու միաբաններու կրած անցքը տեսնելով Մխիթարայ դեռ կենդանի աշակերտներն եւ ուրիշ միաբաններ, ինքնակամ կը հետեւին աքսորեալներուն՝ Կ. Պոլիս երթալու պատրուակաւ ի Դրիեստ ցամաք ելլելով:

1773 Մայիս 19ին կը հալածուին ի մի, երբ կը յիշեն երիցագոյններէն ոմանք թէ Մխիթարայ ծրագրին մէջ կար Տրիեստ վանատուն մը բանալ, եւ կ՛որոշեն միաբան այս միտքը իրագործել: Ատ այս նաեւ Տրիեստի եպիսկոպոսն եւ քաղաքապետությիւնը կը շնորհեն իրենց ամեն դիւրությիւն":

[104] Ormanian, *Azgapatum*, vol. 3, §2108, 3077. "Այլ թէ ինչ էին Մխիթարի *կարգերը* եւ ինչ Մելքոնեանի *նորածեւերություններ* պիտի չկարենանք ճշդել, որոշ տեղեկություններ չգտնելով. Միայն յայտնի կը տեսնուի թէ Մխիթարի կարգերը հաստատուն եւ որոշ հիմերու վրայ դրուած չեն եղեր, եւ միայն իւր անձնական հեղինակությամբ յառաջացեր է իւր ձեռնարկը:" "However, as to what Mkhit'ar's *canons* were and what the *reforms* of Melkonian, we are unable to clarify on account of not obtaining a certain type of documentation. The only thing that is clear is that the *canons* of Mkhit'ar did not seem to be placed on firm foundations and his endeavors were pursued on the basis of personal authority."

canons of Saint Anthony the Abbot, an Egyptian third century C.E. monk, who is widely revered as the founder of Christian monasticism. Referring to how, in 1705, Mkhit'ar sent two of his most trusted disciples, Father Eghia and Father Hovhannes, to Rome to present the constitution he had drafted for his new order, Torossian writes, "He gave them a brief constitution which he had created by drawing from the canons and life of Saint Anthony, as well as from the advices of other spiritual fathers..."[105]

This constitution, whatever it may have looked like, does not seem to have been approved by the Cardinals for reasons that Torossian attributes to the libelous attacks against Mkhit'ar, then made by his detractors who were legion in Istanbul. Years later, Mkhit'ar was persuaded by the Holy See to choose a constitution from one among three monastic orders that were then recognized by Rome: Saint Basil, Saint Augustine, or Saint Benedict. He chose the last one, and the new constitution he drafted was finally ratified by Rome, in 1712.[106] Interestingly, as Matteos Evdokiats'i notes in his *Chronicle* of the island, it was only after the approval of this constitution that Mkhit'ar came to be called Abbot: "before this, he was called preeminent father and, sometimes, the director."[107] The evidence presented above demonstrates that the Mkhit'arist congregation, contrary to Ormanian's assertion, did *indeed* have a firm constitution during Mkhit'ar's tenure as Abbot and that this constitution was recognized by Rome as early as 1712. The evidence also reinforces Akinian's suggestion that, sometime after coming to power in 1750 (and probably around 1755), Melkonian decided to introduce significant alterations to

---

[105] Hovhannes V. Torosian, *Vark' Mkhit'aray Abbayi Sebastats'voy* [The Life of Abbot Mkhit'ar or Sebasea] (Venice: San Lazzaro, 1901), 188. "...տուալ անոնց ձեռքը համառօտ սահմանադրութիւնը գոր յօրինէր էր քաղելով Ս. Անտոն Աբբայի վարքէն եւ կանոններէն և ուրիշ հոգևոր հարց խրատներէն."

[106] Matteos Evdokiats'i, *Ժամանակագրութիւն սրբազան կարգի միաձանգն Հայոց...* [Chronicle of the Sacred Congregation of Armenian Monks...], folio 161. "Զի որպէս 'ի վերանդր պատմեցաք, էզոյց մեզ սրբազան ժողովն, թէ այլ նոր կանոն ոչ է մարթ յաւելուլ այլ զմինն 'ի կանոնաց հարց սրբ[ո]ց, որք 'ի կաթողիկէ եկեղեցւոջ վաղ ժամանակօք հանդիսացեալ են, պարտ է ընտրել եւ ընդ ամէն զհամահալանութի[ւն] նորոգեալ սահմանադրութե[ան] կրօնին ընդունիլ: Եւ միաբանութի[ւն] մեր զկանոն սրբոյ հօրն մերոյ բենեդիքդոսի ընտրեաց:" "For as we have retold above, the Holy Congregation informed us that it was not possible to add a new canon to the canons of the Holy Fathers that existed in the Catholic church from ancient times and that it was necessary to choose from among those approved by the Holy See. [And accordingly] our congregation chose the canons of our holy father Benedict." The standard work on Monasticism is C.H. Lawrence, *Medieval Monasticism: Forms of Religious Life in Western Europe in the Middle Ages,* second edition (London and New York: Longman, 1989). See chapter two on the order of Saint Benedict.

[107] Ibid. 161. "Եւ որպէս վերագոյնդ գրեցաք. յետ այսու օրինակալ միաբանութե[անս] մերոյ հալանութե[ամբ] ս[ր]բ[ո]յ պապին զկանոն եւ զսահմանադրութիւն աբբայական կարգի ընդունելոյ, ապա մխիթար րաբունին, որ գլուխն էր կարգիս մերոյ. ապպայհայր անուանեցաւ. ք[ան]զի յառաջ՝ հայր վեհագոյն կոչիւր եւ երբեմն վերատեսուչ:" "As we wrote above, after our congregation accepted the canons of the constitution approved by the Holy Pope, Mkhit'ar the master who was the leader of our order was called Abbot. Before this, he was called his preeminence and sometimes the director."

this original constitution, and he did so without consulting with his fellow monks (as was the custom during Mkhit'ar's rule), and that this brazen act was among the important catalysts that drove Babikian and Gasparian headlong towards the Great Schism. This is surely the most important insight to be gained from Akinian's account to which we shall return later. On the face of it, this seems like a reasonable account of what caused discontent leading eventually to a schism, even if, at the time, Akinian appeared to lack any documentation for it. As we will see in more detail below, this is a view that is supported by the ample evidence from the archives and is also backed up by Casanova's account discussed earlier. At least one formidable scholar, however, has turned this argument on its head and rejected its import.

In a crucial section titled "The Cause of the Division," in volume three of his magisterial history *Azgapatum* [National History], Maghakia Ormanian turns his erudition and unrivaled expertise as a scholar of the Armenian church to the thorny question as to why the event that we have referred to as the "Great Schism" of San Lazzaro occurred. Here, Ormanian rejects Akinian's and Teodorian's earlier views that "changes to the canons established by Mkhitar" and Melkonian's "use of the sum of money raised by the Indian-Armenians" were the causes of the conflict. "Certainly, however," he writes "the question of the [future] direction [*ughghut'iwn*] [of the Congregation] that was capable of giving rise to such great consequences was more important than these types of *internal, specific, and accidental* factors. Otherwise, we would have to ascribe equally to both sides the stain of being narrow-minded and shallow."[108]

In other words, for Ormanian, we must ignore a conflict of personal ambitions between Melkonian and Gasparian/Babikian, as well as disagreements over the constitution of the congregation as "accidental" or "specific" factors in the Great Schism of 1773 and focus instead on "deeper" and more consequential factors. These had to do with the rival and irreconcilable theological positions between the proponents of the two factions that later became fully manifest in the policies pursued by Azarian and his heirs. Although Ormanian is never explicit in his reasoning regarding the question of the "cause" (պատճառը) of the schism or division, he seems to be implying that distinctions between various factors leading to or causing the "event," or եղելութիւն, of the Great Schism ought be be kept in mind.

---

[108] Ormanian, *Azgapatum*, vol. 3, §2109, 3078. "Եղելութեան հոգին բացատրել եւ արժէքը գնահատել պէտք էր ստուգիւ, բայց հարկ եղած ներքին տեղեկութիւնները չունինք, թէպէտ երեւոյթներն ալ կրնան նոյն նպատակին նպաստել: Մխիթարի սահմանած կանոններուն մէջ փոփոխութիւններ մուծանելու խնդիրին հետ, Հնդկահնոց մէջ հաւաքուած գումարի մը կիրառութեան խնդիրը կամ կերպն ալ խօսուած է․ բայց անշուշտ, այս տեսակ պարագաներէն աւելի, կշիռ ունեցած է ուղղութեան խնդիրը, որ կրցած է այդչափի ծանր հետեւանքի պատճառ ըլլալ: Նպա թէ ոչ հարկ կ'ըլլար երկու կողմերուն վրայ ալ փոքրոգութեան եւ անլրջութեան արատը քսել:" Emphasis added.

According to this view, the ultimate cause or factor that motivated the founders of the Trieste and Vienna orders to abandon the mother convent was connected to the question of the future direction, or ուղղութիւն, of the Congregation founded by Mkhit'ar. He explains this quite explicitly:

> And the question of direction of course consisted of the orientation to be adopted towards the Roman church, on the one hand, and the Armenian one, on the other. These together constituted the crux of the matter at the time. The personal path of Mkhit'ar had its Armenophiliac [*hayaser*] aspects, even though it was swayed by the enigmatic idea of forming a luminous sense of being Armenian within the Roman church. And if this, too, were the intent of Melkonian, his decision [during the Great Schism] would have been inclined in that direction. Following the division [of the order], that concealed spirit of Mkhit'ar indeed continued in Venice, and Armenian studies received a new impulse there. It is worth noting here that three key and meritorious names, Fathers Vrtanes Askerian, Step'annos Agonts', and Mikayel Chamchian, all vardapets, stayed on Melkonian's side. On the contrary, the side that remained with Babikian day-by-day took its loyalty to the Roman Church to extremes and not only did not sustain Armenianness but also by doctrine and teaching, procedure and orientation, even in its liturgical forms and religious garb encouraged the Roman orientation and followed it.[109]

In other words, for Ormanian, Melkonian's explicit and *subsequent* support of Mkhit'ar's position of not abandoning the Armenian Church and its traditions, but finding an ecumenical solution and reconciling the Armenian tradition represented by its Apostolic Church with the positions of the Church of Rome, was the guiding "spirit" (ոգի) that propelled forward the Great Schism. In this view, Babikian and Gasparian were only advocates of a pro-Roman position that saw the Armenian Church and its positions as nothing short of heretical or schismatic. To be sure, the history of the Vienna branch of the Mkhit'arists is far from being reducible to pro-Roman and anti-Armenian church zeal, and Ormanian is aware

---

[109] Ibid., § 2109, 3079-78. "Իսկ ուղղութեան խնդիրն ալ անշուշտ մէկ կողմէն Հռոմէականութեան եւ միւս կողմէն Հայութեան հանդէպ պահուելիք կերպերն էին, որ խնդրոյ ձիւթ կը կազմէին այդ ատեն։ Մխիթարի անձնական ուղղութիւնը՝ հայասէր կողմն ունէր, բայց Հռոմէականութեան մէջ փայլուն հայութիւն մը կազմելու առեղծուածային գաղափարով կ՚օրօրուէր։ Եթէ այս էր Մելքոնեանի ալ դիտումը, հարկաւ որոշումը՚անոր նպաստաւոր կողմն պէտք է հակի։ Բաժանումէն ետքն ալ Մխիթարի այդ ծածուկ ոգին շարունակեց Վենետիկի մէջ, եւ հայկական ուսումներ նոր զարկ առին։ Նշանակութեան արժանի է, որ երեք գլխաւոր եւ արդիւնաւոր անուններ, ինչպէս են՝ Վրթանէս Ասկերեան, Ստեփաննոս Ագոնց եւ Միքայէլ Չամչեան վարդապետները, Մելքոնեան բաժնին մէջ մնացին։ Ընդհակառակն Պապիկեանի հետ եղող բաժինը օր աւուր իրեն Հռոմէականութիւնը դէպի լատինամոլութիւն տարաւ, եւ ոչ միայն հայկականութիւնը չպարզրացոյց, այլեւ վարդապետութեամբ եւ ուսումամբ, ընթացքով եւ ուղղութեամբ, եւ նոյնիսկ ծիսական կերպերով եւ զգեստներով՝ կատարեալ լատինութեան կողմը քաջալերեց եւ անոր հետեւեցաւ։"

of this. For nearly a whole century from the Great Schism onwards, Ormanian explains, "the reigning element in the [Trieste/Vienna] branch was entirely an overzealous form of Roman Catholicism, all the while when the branch that stayed with Melkonian was subject to persecutions from the overzealous ones on account of their desires for supporting Armenophilia and intimacy with the ways of the Armenian Church."[110]

What is interesting to note about the above interpretation is that it does not vitiate Akinian's claims that alterations to the constitution and lack of transparency in rule (what Casanova identified as Melkonian's "despotism" or "tyrannical" tendencies) were factors involved in paving the path for the Great Schism. Nor for that matter does Ormanian deny that Melkonian's misuse of funds coming from India as proposed by Casanova, Teodorian, and later Zavrian played any role in the separation of the two branches. Rather, the argument he advances seems to accept these as "internal," "specific," and "accidental," in short *superficial*, factors but asserts that the *real* cause or "spirit" was connected to the theological/ideological positions of the rival factions. Ormanian, of course, is cautious in articulating this view, knowing full well that the requisite "internal information" (presumably in the form of documentation left behind by the actors or a chronicle of the events kept by an eyewitness) was either not available or, if it was, was not at his disposal. It bears remembering here that the third volume of his *Azgapatum* was written in Istanbul in 1914 with no direct or even indirect access to the archives in Venice, Trieste, Rome, or Vienna. Any admirer of Ormanian will be quick to note that his documentation, in these sections at least, is rather sketchy and thin and consists of a handful of secondary sources, including Akinian and Teodorian. Eschewing earlier explanations and with no additional archival documentation, Ormanian thus privileges deep-seated theological and ideological motives concerning the future "direction of the congregation." What is one to make of such interpretation, and how does this view stand up to the new documentation not available either to Ormanian or to Akinian but since unearthed and at the disposal of scholars?

In an all-too-obscure essay on the "Separation of the Mkhit'arist Congregation" published in 1932, Hovhannes Zavrian finds Ormanian's arguments baseless.

> It seems to me that Ormanian accepts consequence as cause. It is true that the Mkhit'arists of Vienna, from the very first day of their break, showed themselves to be more fervent Catholics than those in Venice, and perhaps that is how they are today as well. However, there is no basis at all for insisting that such abstract and principled issues were causes for the schism. On the contrary, the separation

---

[110] Ibid. "Իրաւ է որ վերջին ժամանակներու մէջ՝ կղեմէս Սիպիլեանի, Ղուկաս Տէրտէրեանի եւ Արսէն Այտընեանի ընդգրկած ուղղութիւնը, քիչ մը փոփոխեց Պապիկեանով սկսած ճիւղին ընթացքը, սակայն մինչեւ այսօր ատենը, որ է ըսել դար մը ամբողջ, բոլորովին նախանձայոյզ հռոմէադաւանութիւնն էր այդ ճիւղին մէջ տիրողը, մինչ Մելքոնեանի հետ եղած ճիւղը Հայասիրութեան իղձերուն եւ հայամօտութեան կերպերուն համար, նախանձայոյզներուն հալածանքներուն եւ պապութեան խստութիւններուն ենթարկուեցաւ։"

happened because of causes that flowed from much more simple and human passions and desires.¹¹¹

True, Ormanian was more than likely projecting backwards the climate of opinion produced by the Catholic-Apostolic sectarian strife that characterized the public sphere of Armenian debate and discussion in the Armenian periodical press of Istanbul, Izmir, and elsewhere for much of the nineteenth century. His interpretation may be seen, in other words, as an instance of what historians refer to as "presentism," the tendency of anachronistically reading backwards the ideas, values, and assumptions from one's present as explanatory factors back into the past. Coming of age as a former Catholic turned prominent member and even patriarch of the Armenian church in Istanbul during the second half of the nineteenth century, it is hard not to see Ormanian's "present" as deeply shaped by the polemics and diatribes for and against the Mkhit'arist Congregation in Venice launched by the libelous book *Il Mechitarista di San Lazzaro* that appeared in Istanbul 1852. This work accused the Venice branch of the Mkhit'arist Congregation as being crypto-heretics all-too-willing to defend the "schismatic" positions of the Armenian Church, while pretending to be Catholics. Some, at the time, believed that members of the Vienna branch of the Mkhit'arists, and Abbot Azarian in particular, were behind the release of this anonymous libel.¹¹² It is likely then that, as Zavrian avers in his 1931 essay, Ormanian was merely projecting backwards, in an anachronistic fashion, theological/ideological differences between the two orders that had emerged *after* the schism into the motivations and intentions of Gasparian and Babikian, in 1773. If this were the case, and theological or sectarian motivations were not the cause of the Great Schism but its consequence, what then are we to make of the cause? Were we to dismiss the lofty-sounding, theological/sectarian concerns over the two rival "directions" of the order Mkhit'ar left behind as the real causes of the rift, would we then be compelled, as Ormanian asserts in the passage quoted earlier, "to ascribe equally to both sides the stain of being narrow-minded and shallow"? Zavrian seems to think so when he concludes, "the separation happened because of causes that flowed from much more simple and human passions and desires."

Curiously, the scholar looking for "motives" in the Inquisitori di Stato Archives or the State Archives of Vienna and Trieste is likely to conclude, as Ormanian shuddered to think and as Zavrian claimed, that the Great Schism indeed resulted from petty issues involving personality conflicts, "human passions and desires," as it were. Indeed, the extant

---

¹¹¹ Zavrian, "Mkhit'arian Miabanut'ean Bazhanumě," 97. "Ինձ թւում է որ Օրմանեանը Հետեւանքը ընդունում է որպես պատճառ: Ճշմարիտ է, որ Վիեննայի Մխիթարեանները, անքատման Հէնց առաջին օրերից սկսած, իրենց ցոյց տուին աւելի քերմ կաթոլիկներ, քան վենետիկցիները եւ, թերեւս, Հիմա էլ նոյնն են, սակայն, ոչ մի Հիմք չկայ պնդելու, թէ պառակտման պատճառ են ծառայել այդպիսի վերացական-սկզբունքային խնդիրներ: Ընդհակառակը, բաժանումը առաջ եկաւ շատ աւելի պարզ եւ մարդկային կրքերից եւ ցանկութիւններից բխող պատճառներից:"

¹¹² Ibid., 77.

documentation does reinforce this view that a struggle of "big egos" was at play, pitting a young and domineering Melkonian against the "temperamentally volatile" Gasparian and the ever-more ambitious Babikian, who hailed from a wealthy family with connections to India and London. As with virtually everything in human history, the peculiarities of individual agency (including power struggles) were also no doubt involved here. Contrary to Ormanian, however, *no* evidence exists that theological or sectarian positions played *any* significant and known role. Is there no other explanation at hand that provides more nuance and complexity to the "human passions" account provided by Zavrian? What light do the archives have to shed on this vexed question? One unlikely source, overlooked by Zavrian, Ormanian, Curiel, and virtually everyone else who has delved into the matter, is quite promising in this respect.

This source is a 14-page report written by Abbot Melkonian on 2 October 1773 and sent to the College of Cardinals, comprising the *De Propaganda Fide* that we have already quoted from above. In it, Melkonian explains that despite informing the Papal nuncio in Venice from time to time of developments on the island connected to the events of 1773, he will take the opportunity in this report of personally discharging his duty and representing "the origin, progress, and the felicitous conclusion of the affair [the Great Schism] to the profit of our congregation."[113] What is stunningly insightful in this report is the candor with which the Abbot addresses the genesis of the conflict, which he attributes overwhelmingly to Gasparian and the "agitated youth" (*fomentati giovani*) who followed him and began demanding new measures (*nuovi provedimenti*) regarding the island's constitution. After reluctantly agreeing to call a chapter, or assembly (*capitolo*), of the capitular, or senior monks, to discuss these reforms, Melkonian quickly realized the dangerous "threats" that "such an aggression" posed to the security and order of his island. He reports,

> In the first congresses, it seemed that the aim of all the capitular monks was to promote the disciplinary rules [within the congregation] with the exact observance of the constitutions, and other regulations, which were believed necessary on the basis of the constitutions for the domestic and internal governance of the same community. However, after some preliminary sessions, Father Minas with his colleagues came little by little to place into question the very validity of the venerable decree from this Sacred Congregation [i.e., the *De Propaganda Fide*] issued on December 4, 1762.[114]

---

[113] ASPF, SC. Armeni 17, report of Stefano di Melchiore to the College of Cardinals, 2 October 1773, folio 441. "...vengo personalmente in adempimento del mio dovere a rappresentarne l'origine, il progresso, ed il felice esito della cosa con profitto della nostra Congregazione."

[114] Ibid. "Ne' primi Congressi sembrava, che la mira di tutti i Capitolari fosse diretta a promuovere la Regolare Disciplina coll'esatta osservanza della Constituzioni, ed altri Regolamenti, che fossero creduti necessari sulla norma delle Costituzioni medesime per l'interno domestico governo della Comunità, ma dopo alcune preliminari sessioni, a poco venne il P. Minas co suoi Collegati a proporre in questione la validità del venerabile decreto da codesta S. Congregazione emanato a li 4 Dicembre 1762."

The reference here, of course, is to the *De Propaganda Fide's* and Pope Clement XIII's official recognition of Melkonian as the Congregation's Abbot, and the Papal ratification of the post as a lifetime appointment. It appears that, up till then, Rome had not ratified Melkonian's initial election as Abbot by the chapter of capitular monks on 9 April 1750, and neither was the length of Melkonian's tenure spelled out. Moreover, all the circumstantial evidence also points in the direction that Melkonian had altered the monastery's constitution from the time of its final ratification, in 1712, under Mkhit'ar and submitted the revised version to the Cardinals at the *De Propaganda Fide*, sometime in the mid-1750s, as Akinian suggests. Under Mkhit'ar's tenure, the Abbot, or "superior" (մեծաւոր), as the holder of the office was also known at the time, was expected to rule for life. It seems unclear whether following Mkhitar's death his successor was also expected to hold the office for life or for four years. What is clear is that beginning in the late-1750s, when Melkonian began making a formal request to the College of Cardinals for official recognition of his election, he appears to have submitted a new constitution asking for significant expansions to his power as Abbot as well as a lifetime appointment. This had created considerable tension on the island already at the time, as a confidential missive from Istanbul written to the Cardinals on 15 November 1759 by Hakobos Chamchian makes abundantly clear.[115] In this heretofore-unknown letter, the brother of the famous historian outlines all the traits that disqualified Melkonian from holding such an important office (his "contempt for his inferiors," his will to power, and his unstable tempreramment being key in his estimate).[116] He also warns the Cardinals that Melkonian's "intent is to sit permanently as the superior [i.e., Abbot] with *absolute* powers, independent of the councilors"[117] and states that eleven of nineteen monks were sternly opposed to his lifetime appointment and would "exit the congregation" if Melkonian's tenure

---

[115] ASPF, Congregazioni Particolare (CP) 127, 95r-96r. I am grateful to Cesare Santus for sharing copies of this precious document with me.

[116] Ibid., 95r. "Սա թէպէտ խոհեմ ոք է յիրս ումանս, բայց անընդունակ է մեծաւորական կառավարութ[ան]. եւ այս յայտ է այսու, զի ահա տասն ամ է եւ հ[անա]պ[ա]զ կռիւ է 'ի միաբանութ[եան]. եւ խէթ ականրկումն է ն[ո]ր[ա] առ ումանս ստորադրեալս իւր եւ ն[ո]ց[ա] առ նա." "Though, he is a prudent man in certain things, he is also incompetent at the task of governing as a superior. And this is evident on account [of the fact that] for the past ten years there have been continuous quarrels in the congregation and also on account of the contempt with which he regards some who are his inferiors and of theirs towards him."

[117] Ibid. "Այլ դիտումն ն[ո]ր[ա] այս է, զի ինքն ասդի մշտնջենաւոր մեծաւոր, եւ այն բացարձակ իշխանութեամբ անկախ յառաքակայից:" Emphasis added.

and powers were extended.[118] At least four had apparently already indicated this in their talks. Interestingly, neither Gasparian's nor Babikian's names are included in the list of dissentors whom Chamchian worried would exit the congregation and, thereby, destroy the tiny order Mkhit'ar had left to his heirs. At any rate, Chamchian's letter now seems prophetic in its dire warnings that Melkonian's consolidation of power might destroy the congregation.

For reasons that need to be examined further, the Cardinals ignored Chamchian's grave concerns. On 4 December 1762, they extended a formal recognition of Melkonian as Abbot for life, a recognition that was simultasneously made by the Pope Clement XIII. This opened the way for Chamchian's prophecy to unfold. The "agitated youth" led by Gasparian and Babikian, on whom Melkonian blamed the tumult on his island, was only carrying out the dissent and opposition to Melkonian's drive for absolute power that was already present by the time Chamchian wrote his alarming letter. It appears, thus, that this "youth" did not appreciate being left out of the decision-making process and were pressing to have a place at the table. Their ringleaders, Gasparian and Babikian, succeeded "to win over the majority of the assembled monks [capitolari] to the belief that the esteemed decree [of December 1762] was apocryphal [not authentic] and without value." Melkonian vigorously opposed this "bold aggression" (i.e., their decision to question the validity of the 1762 decree from Rome) and was even more astonished when Gasparian and Babikian went further in revealing the real intent and objectives behind their calling for a chapter or assembly:

> However, once they saw their hatched plan dissipated, they raised a more serious raucous. And engrossed more and more in their schemes, they came up with a new project, that in the future everyone was to be free to express their own judgment [*sentimento*] on the propositions brought before the chapter or assembly [*capitolo*], and after having the subject discussed they had to put everything to a vote [*ballottazione*], and everyone had to take an oath to obey, once and for all, what had been decided by a secret vote of the majority of the participants of the chapter. In this I also saw the malignant intention and pernicious consequences

---

[118] Ibid., 95v. "Արդ՝ գիտեմ հաստատեալ, որ թէ մշտնջենաւորաբար եւ բացարձակ իշխանութ[եամբ] լիցի մեծաւորն մեր, եւ զայս իշխանութի[ւն] ունիցի ներկայ մեծաւորս մեր, գրեթէ այնք վերոգրե[ա]լ մետասան անձինք չկամեցող զմշտնջենաւորութի[ւն] մեծաւորին, ելանեն 'ի միաբանութ[են]է, յորժամ նորոգեցցի ուխտն 'ի հաստատելն վեմութե[ան] ձերոյ զկացութիւն մեր։ Որպէս արդէն չորք անձինք Հ. Ներսէս, Հ. Անտոն, Հ Մեսրով, եւ Հ.Անանեա. խոսով են ընդ մեծաւորին. եւ սպասեն ելանել 'ի վանքէն եթէ մեծաւորն մշտնջենաւոր լիցի: Նաեւ երկուք կան պաշվալով, որ Հ. Մարկոս, եւ Հ, Սահակ. որք այսմ սպասեն. եւ 'ի ներկայ մեծաւորես գրեթէ անդր ապսորեալ կան:"

"Now, I know with certainty that if our superior were to be appointed for lifetime and with absolute powers, nearly all of the eleven persons who do not wish to have the office of the superior be a lifetime appointment, will exit the congregation..."

that would have ensued, so I forcibly opposed myself to this vote and I did not want to consent to any oath whatsoever.[119]

The evidence provided in this report suggests that what galvanized opposition to Melkonian had nothing or little to do with the future "orientation" of the congregation or even with theological issues, as Ormanian argued a hundred years ago. Neither Melkonian in this report nor any other known eyewitness source, in the immediate aftermath of the Great Schism, attributes the conflict to theological or doctrinal differences among the parties. Such differences, as we have seen, emerged later and are largely early-nineteenth-century phenomena reflecting the sectarian clashes that broke out in Istanbul and Izmir. Rather, the evidence from Melkonian's report indicates that something other than either theological or sectarian disagreements (as Ormanian contends) or personality conflicts (Zavrian and Teodorian maintain) was at the genesis of the conflict. Constitutionalism, or a struggle for more representative governance, appears to have been at the crux of the Great Schism.

Here, it is important to note that while most monastic orders in the Catholic world had chapters (capitoli) where capitular monks discussed and debated issues important to their congregation, it was generally the rule that superiors or abbots had the final and decisive say in making decisions that were binding on all. The Mkhit'arist order in San Lazzaro was no different in this respect. It followed the Rules of Saint Benedict, which were quite strict and inculcated an "uncompromising doctrine" where "the personality of the abbot was the linchpin of the monastic community." According to C. H. Lawrence, for instance, the abbot could "appoint and dismiss subordinates, allocate punishments, and direct the relations of the monastery with the outside world as he thinks best." However, he was "urged by the Rule to take the advice of the brethren before taking policy decisions, but he is not bound by it. Constitutionally then, St. Benedict's monastery is a paternal autocracy tempered by the obligation to listen to advice."[120] In light of the stipulation that abbots were expected and often did take into account the counsel of their brethren, Babikian and Gasparian's demands were brazen but not entirely out of bounds. At any rate, this was not the novel element in their request. What was new and unusual with their demands (at least as these are articulated in Melkonian's own words) in 1773 was the individualistic stance, namely the idea that "everyone was to be free to express their own judgment [sentimento] on the propositions brought before the chapter or assembly [capitolo]." This focus on individual

---

[119] Ibid., folios 442v and 442r. "Vedendo però essi dissiparsi la concepita idea suscitarono più gravi tumulti, ed impegnati sempre più ne divisamenti loro, uscirono con un nuovo progetto, e fu, che in avvenire fosse libero a chiunque dire il proprio sentimento sulle proposizioni, che fossero fatte in Capitolo, e che dopo aver discussa la materia, dovesse porsi alla ballottazione, ed ognuno prima di tutto, ed una volta per sempre prestasse solenne giuramento di ubbidire a ciò, che venisse deciso per voti secreti dalla maggior parte de' Capitolari. In ciò pur vidi la maligna intenzione, ed i pericoli effetti, che indi ne sarebbero occorsi: ed a questo pure con forza di ragione risolutamente m'opposi né volli accordare, che si facesse giuramento di sorte alcuna."

[120] Lawrence, *Medieval Monasticism*, 29.

rights of freedom of expression would strike most observers now or then as being nothing short of revolutionary. It also raises the question of whether such an idea might have been a sign of the times, a kind of monastic articulation of a revolutionary way of thinking that would only make itself evident less than two decades later. Melkonian's reaction to these ideas as "malignant" in intention and "pernicious" in consequence would have been quite natural to his contemporaries, even as they strike us today as commonsensical. How precisely the monks opposed to Melkonian came up with such notions and whether these ideas had anything to do with the climate of representative governance that was in the air during the decade or two before the Great French Revolution is a matter that deserves deeper study.[121]

Conclusion

Despite its significance to early modern Armenian history and notwithstanding the large corpus of scholarship devoted to Abbot Mkhit'ar and the monastic order he founded, the Great Schism of 1773 leading to the bifurcation of the congregation into two, at times, rival orders remains virtually unstudied and unknown. It occupies a spectral presence on the margins of Mkhit'arist scholarship and historical memory, almost like a non-event. Mkhit'arist monks and scholars who have contributed the lion's share of the writings on their own history appear to have deliberately avoided this pivotal turning point in their island's past. For them, Father Step'an Sarian's cryptic observation of 1901, quoted at the outset of this essay, that "dark clouds passed over the arches of San Lazzaro" seems to have sufficed. Even a scholar as talented as Leo, who has correctly identified the events of 1773 as constituting nothing less than a "revolution," has ambiguously referred to their genesis as "internal opposition" and "disorder."[122] For Mkhit'arist monks, the obfuscation and silence that have fallen on this issue are no doubt symptoms of a cautious policy of avoiding a painful chapter of their order's rich history. This may be motivated by a desire not to open old wounds and openly discuss an episode that might appear to some as potentially embarrassing. For scholars outside the congregation, unfamiliarity with and inaccessibility of the archival sources surely would figure among the important reasons why the Great Schism has, until now, remained relegated to obscurity. The result is that only a handful of studies, often by scholars who are outliers to Mkhit'arist history, have broached the topic. Their works, however, have been long forgotten and sidelined.

Relying on previously unknown or little-used archival documentation stored in the state archives of Venice, Trieste, Vienna, and Rome, as well as the memoirs and correspondence of Casanova, this essay has critically reconstructed the events that led to the Great Schism of 1773. My microhistorical study has illuminated two larger issues that are of vital importance to scholars interested in early modern Armenian history. First, the Great Schism, as I have argued above, sheds important light on the previously shadowy and unstudied history of

---

[121] The role of constitutional governance in monastic orders is not covered in Francis Oakley's *The Conciliarist Tradition: Constitutionalism in the Catholic Church, 1300-1870* (Oxford: Oxford University Press, 2003).

[122] Leo, "Mkhit'areanner," 521 and 510. See Also Yerits'eants', *Venetiki Mkhit'areank'*, 42-4.

the Armenian diaspora of Trieste during the last quarter of the eighteenth century. In this connection, I have argued that the Habsburg authorities were motivated to grant privileges to their Armenian subjects, including the expelled monks, as a result of mercantile factors connected to the development of their free port at the expense of neighboring ports such as Venice. Second, my close reading of the events of 1773 also sheds light on the identity of the Mkhit'arist congregation and the breakdown of authority following the election of Mkhit'ar's successor Abbot Step'annos Melkonian. To this end, in the course of examining the genesis of the Great Schism, I have argued that the core issues that led to the separation of the order into two branches had little, if anything, to do with theological factors or different religious world views as Ormanian had suggested. On the contrary, all the evidence points in the direction that at the genesis of the Great Schism was a struggle between Melkonian and Gasparian/Babikian over the constitution of the order and the legitimacy of what Melkonian might have done with the constitution left behind by Abbot Mkhit'ar. In short, a struggle over constitutionalism and representative monastic governance, as well as concerns over Melkonian's mismanagement of funds were the two principal causes that gave rise to the separation of the congregation in 1773. A more detailed and comprehensive study of the matter will surely further clarify the issues in ways beyond the scope of this preliminary exploration.

THE ARMENIAN OIKOUMENE IN THE SIXTEENTH CENTURY: DARK AGE OR ERA OF TRANSITION?

S. Peter Cowe

The Problem

With a genealogy reverting to Petrarch and an evolution spanning the next several centuries applied to a concertina of timeframes, we are familiar with the term Dark Age largely from European history. Generally conceived as originating in the fall of the Roman Empire in the West, the term Dark Age has labelled respectively the period up to the Renaissance or Reformation, while, in more recent iterations, the epoch extending to the rise of Charlemagne and the inauguration of the Carolingian Renaissance. This account would accordingly dramatize the role of Irish monks on the Atlantic seaboard, saving civilization by a thread through perpetuating Latin manuscripts by a sputtering candlelight.[1] However, more recent scholarship has significantly revised this narrative after uncovering the high degree of contact those communities had with centers around the Mediterranean basin, suggesting far more activity and continuity than had previously been considered.[2]

How do these developments relate to the situation in Armenia? In his chapter edited by Dédéyan, Kouymjian states, "Les XVᵉ et XVIᵉ siècles peuvent être appelés l'âge obcur de l'histoire arménienne,"[3] a view he repeats more unequivocally in his chapter in a parallel volume edited by Hovannisian, remarking "The fifteenth and sixteenth centuries are the dark

---

[1] See K. Clark, *Civilization: A Personal View* (London: British Broadcasting Corporation, 1969), 7-14.

[2] C.A. Snyder, *An Age of Tyrants: Britian and the Britons A.D. 400-600* (University Park: Pennsylvania State University Press, 1998), xiii-xiv.

[3] G. Dédéyan, *Histoire des Arméniens* (Toulouse: Éditions Privat, 1982), 341.

ages of Armenian history."⁴ What are the metrics for its delineation within that category? Significantly, the main architect of the concept, Cardinal Caesar Baronius, himself a denizen of the 16th century, identified the relative abundance or dearth of written records as the determinative criterion, leading him to postulate two dark ages in Western Europe, preceding and following the Carolingian Age, both characterized by a fall in manuscript production suggestive of a corresponding decline in intellectual activity. In keeping with this, it has often been observed that no significant historian exists between T'ovma Mecopec'i (d. 1446) and Aṙak'el Dawrižec'i, who completed his work in 1660.⁵ This judgment, however, ignores the role of figures like Grigor Daranałc'i, whose chronicle, though completed in 1634, covers events of the second half of the 16th century.⁶ This coverage is then supplemented by continuators of earlier chronicles like that of Samuēl Anec'i⁷ and contemporary local chronicles, such as that of Yovhannēs Arčišec'i and Yovhanisik Carec'i,⁸ and a series of historical poems tabulating the impact of the Ottoman-Safavid hostilities on important urban centers.⁹

More broadly, this paper will argue for a reexamination of the above perspective both regarding sources and the conceptualization of the period and Armenian agency exercised within it. In what follows, I will review a variety of contemporary literary works of different genres, including martyrologies, translations, and works of fiction with significance for history. I suggest that the insight those multiple sources afford into the larger scale developments of the time and the concomitant diversification of Armenian sociopolitical and economic roles, after the redrawing of the Armenian Plateau, offers us a more rounded view of events than the previous historiographical tradition. Thereby, they help us document the movement of Armenians from the Plateau to a diasporic setting and from a rural landscape to an increasingly urban ambience, with a parallel transition in leadership from aristocracy to the middle class and inchoately in authorship from the celibate monastic hierarchy to married clergy serving the communities, as well as laymen positioned more closely to events to capture the process and implications behind this metamorphosis.

---

⁴ R.G. Hovannisian, *The Armenian People from Ancient to Modern Times*, vol. 2 (New York: St. Martin's Press, 1997), 1.

⁵ See D. Kouymjian, "Dated Armenian Manuscripts as a Statistical Tool for Armenian History," in *Medieval Armenian Culture*, ed. T.J. Samuelian and M.E. Stone (Chico, CA: Scholars Press, 1984), 425; R.G. Hovannisian, *The Armenian People from Ancient to Modern Times*, vol. 2 (New York: St. Martin's Press, 1997), 1.

⁶ For this work, see M. Nšanean, *Žamanakagrut'iwn Grigor vardapeti Kamaxec'woy kam Daranałc'woy* [Chronicle of the vardapet Grigor Kamaxec'i or Daranałc'i] (Jerusalem: St. James Press, 1915).

⁷ See K. Mat'evosyan, *Samuēl Anec'i ew šarunakołner* [Samuēl Anec'i and Continuators] (Yerevan: Nairi, 2014), 115-357.

⁸ V. A. Hakobyan, *Manr žamanakagrut'yunner XIII-XVIII dd.* [Brief Chronicles 13th-18th cc.], vol. 2 (Yerevan: Armenian Academy of Sciences, 1956), 225-34 (Arčišec'i) and 235-55 (Carec'i).

⁹ For details, see P.M. Xač'atryan, *Hay mijnadaryan patmakan ołber (žd-žē dd.)* [Medieval Armenian Historical Laments] (Yerevan: Armenian Academy of Sciences, 1969), 91-144, 214-49.

Martyrology as a Source for Sixteenth Century Armenian History

Let us begin with martyrology, which until today remains relatively untapped, although for social history it is virtually unsurpassed in an Armenian context.[10] Its ideological concerns are concentrated in the judicial proceedings, the subject's bearing under execution, and details of burial, the almost ubiquitous appearance of a ray of light over the martyr's relics frequently observed and marveled at by non-Christians, serving as a token of divine validation of the act. Consequently, the vignettes the authors afford of the martyrs' life style in this period, their interaction with Muslim neighbors and coworkers, and other circumstantial data on ambience in the dynamically changing conditions cumulatively, permit us to assemble many key elements to reconstruct the social impact of the Ottoman-Safavid conflict over the century. Here, I would like to direct the readers' attention to some of the main works that focus our attention on seminal aspects of contemporary history.

The anonymous martyrology of Xač'atur (1517) provides valuable contextualization for the first Ottoman-Safavid War. The protagonist presumably moved his family to Amid (Diyarbakir) for greater protection, while still in Persian hands, before it was captured by the Ottomans in 1515 and established as the center of an eyalet, which acted as one of the main forward troop bases throughout this period. The significant military presence there provides the main engine determining the course of events. Communication problems most likely between Turkish and Persian with different groups of soldiers in the city sets in motion the process leading to his death.[11] Similarly, Mecparon's narrative of the youth Xətəršay, martyred in Marsovan in 1541, speaks to the Armenian situation after the second bout of warfare (1532-35). The author notes that the lack of current leadership on the part of the local Armenian gentry has acted as a catalyst for large-scale Muslim migration to the region, which had left Armenians as unwitting targets for both sides, during the recent war. Moreover, in describing the status quo, he provides valuable information on the application of the *ghiyār*, the dress code of wearing blue as a feature distinguishing non-Muslims, as well as a prohibition on employing white headgear.[12]

Likewise, other representatives of the genre depict the situation during the fifth Ottoman-Safavid War (1578-90), such as Siměon Tigranakertc'i's treatment of the martyrdom of the deacon Margarē and Šahrəman of 1580, which complements Yovanisik Carec'is account. The latter discusses the complex issue of Safavid succession, tracing this from the last years of Shah Tahmasp I's reign (1572-76) to the short reign of his son Isma'il II (1576-77), that of

---

[10] A similar perspective is provided by saints' lives in other traditions, such the Byzantine, but there are comparatively few available in the Armenian.

[11] See Y. Manandean and H. Ač'aŕean (eds.), *Hayoc' nor vkanerə (1155-1843)*[Armenian Neomartyrs 1155-1843] (Vałaršapat: Mother See Press, 1903), 350-52 for the text and D.R. Thomas et al (ed.), *Christian-Muslim Relations*, A *Bibliographical History*, vol. 7 (Leiden, Boston: Brill, 2015), 588-91 for a brief study of its significance.

[12] For the text, see Y. Manandean and H. Ač'aŕean (eds.), *Hayoc' nor vkanerə*, 376-82; and Thomas, *Christian-Muslim Relations*, 668-72, for a brief study of its significance.

the latter's brother Mohammad Khodabanda (1578-87) and the opening years of the latter's son 'Abbas I. In addition, he addresses the contested loyalty of the qizilibash troups, the provincial revolts, and the external threats (Ottoman and Uzbek) that the period of weak central rule had facilitated. In contrast, the martyrology presents the Ottoman perspective on the war Sultan Murat III had broached, profiting from Shah Tahmasp's recent death. It surveys his campaigns against Tiflis, Shirvan, Khoy, and Salmast, all with an important Armenian population, many of whom were taken captive. Moreover, he reports that the devastation inflicted by the cavalry produced a series of poor harvests and widespread inflation. The author also offers useful insights on the integration of the newly acquired territories into the Ottoman provincial administration.[13]

Trajectories of Armenian Manuscript Copying and Monastic Scriptoria

To document statistics of Armenian manuscript copying in the period from the fourteenth to the seventeenth centuries, Kouymjian had undertaken a study of dated codices that displays a noteworthy dip from a production of 832 in the fifteenth century to 627 in the sixteenth, followed by a near doubling of output in the first half of the seventeenth century to reach 1,250.[14] What is even more helpful is the breakdown by decades, illustrating the worst decline occured in the first half of the century, coinciding with some of the harshest Ottoman-Safavid fighting associated with Chaldiran and the War of 1532-55, after which the numbers increased, before dipping slightly again in the last decade, aligning well with the Jelali revolts.[15] These are indeed important considerations, which must clearly be integrated into any synthesis of the period we are to arrive at. But can it be that their prominence in discussion has been disproportionate to their overall weighting? The provenance of most manuscripts would have been monastic scriptoria, which were clearly significantly affected by the unsettled conditions, so much so that the restoration of monastic communities became a key concern of the following century, the initiators of one of the main projects, the Mec Anapat of Siwnikʻ, actually undertaking the journey to Jerusalem to investigate how traditions were being maintained in the monastic heartland in the hopes of emulating them.[16] However, one wonders whether the perch these institutions afforded and the typical mindset they inculcated were adequately calibrated to register the broad trajectory of events and to interpret that against the wider purview of developments across the region and beyond, for the process of negotiating identity for Armenians continued, and the communities were to

---

[13] See Manandean and Ačʻaṙean, 413-20.

[14] It should be noted that seventeenth-century-manuscript copying does not simply mark a return to the status quo ante but reflects the changes of the previous century in terms of new uses of the medium, more specialization on the luxury market, and a more diversified scribal class that features more women practitioners.

[15] D. Kouymjian, "Dated Armenian Manuscripts as a Statistical Tool for Armenian History," in *Medieval Armenian Culture*, eds. T.J. Samuelian and M.E. Stone, (Chico, CA: Scholars Press, 1984), 428-29.

[16] S. P. Cowe, "Church and Diaspora: The Case of the Armenians," in *Cambridge History of Christianity, Vol. 5, Eastern Christianity*, ed. M. Angold, Cambridge: Cambridge University Press, 2006), 437.

express themselves in diverse domains. Hence, we need to interrogate the presuppositions underlying the conceptualization of the era as a dark age, particularly the core issue of whether, in this period, a reduction in manuscript production can be equated with an overall decline in intellectual activity.

The Era in Armenian Historiographical Paradigms

The general tendency in modern historiography has been to prioritize periods of Armenian statehood and thereby to neglect the interstices, perhaps from a false sense of loss of agency.[17] This perspective is particularly betrayed by the widespread currency of the metaphor of the canr luc ("heavy yoke") to describe Armenian conditions under Ottoman and Safavid rule.[18] Indeed, the term features prominently in the heading of one of Kouymjian's chapters cited earlier.[19] Another facet of the same approach is to sketch the epoch in extremely broad strokes, such as portraying Armenian culture with few exceptions as falling "into a period of stagnation that lasted until the nineteenth century."[20] However, Armenians patently do not disappear during those stateless "gaps" or go into hibernation, and that the loss of an autonomous administrative structure does not automatically imply thralldom is suggested, for example, in Mamikonian political rhetoric of the second half of the 5th century.[21] Moreover, it is important to seek out new sources beyond the traditional to help penetrate behind the initial appearance of an undifferentiated tableau to discover nuance and diversity. In this connection, it is an urgent desideratum to edit a collection of manuscript colophons from the sixteenth century, which would assist us in refining our understanding of the period, particularly as they represent a wide geographical spectrum of the Armenian world of the time.

Clearly, the sixteenth century marked an era of unprecedented change on a vast scale, transgressing physical and intellectual boundaries and pursuing technological innovation. Everywhere was marked by dynamic movement, so that scarcely any part of the globe was so isolated as to lie beyond its reach, let alone the Armenian Plateau that had been a thoroughfare between East and West for centuries. Navigation, previously a localized

---

[17] Kouymjian in Hovannisian comments that Armenia in the fifteenth-sixteenth centuries, "is either ignored in standard histories or relegated to a page or two." Kouymjian in R.G. Hovannisian, *The Armenian People*, 1.

[18] *Hay žołovrdi patmut'yun* [History of the Armenian People], Pt.1, eds. B.N. Aṙak'elyan and A.Ṙ. Hovhannisyan (Yerevan: Haypethrat, 1951), 259.

[19] It reads "Sous le joug des Turcomans et des Turcs ottomans (XVe-XVIe siècles)." See G. Dédéyan, *Histoire des Arméniens*, 341.

[20] G. A. Bournoutian, *A Concise History of the Armenian People (From Ancient Times to the Present)* (Costa Mesa, CA: Mazda Publishers, 2003), 183.

[21] Despite the fact their administrative status lacked a legal basis without Persian king's sanction, and that the revolt they were heading was pitted against overwhelming odds, they persisted in their project with strategy and sagacity and were able to make circumstances work in their favor to orchestrate an unprecedented concession from the Persian crown that satisfied their demands.

enterprise in the main,[22] now expanded across the world's oceans with far-reaching economic, military, and political consequences. Whereas, at the outset, the advantages remained the monopoly of Spain and Portugal, by the end of the century, the three other Atlantic powers, England, France, and Holland, were challenging their ascendancy and staking their own claims, which were to achieve fruition over the next century. Similarly, the map of Asia was redrawn with the establishment of three Islamicate empires, the Ottoman, Safavid, and Mughal, Ottoman expansion in particular, not only making inroads into Central Europe but jockeying for position with Venice over control of the Mediterranean and with Portugal over dominance of the Western Indian Ocean. Meanwhile, the Russian principalities united under Grand Prince Ivan as first tsar and began a process of territorial gains from the khanates of the steppes to consolidate communications to the Caspian.

Armenians had established trade communities in several of those regions, over the last centuries, that formed networks of exchange intersecting across the Plateau and interacting with and affecting the life of the main mass of the Armenian population in the homeland. Clearly, the Ottoman-Safavid wars and the subsequent Jelali revolts exacerbated conditions provoked by longer-term forces, such as climate change[23] and the cycle of poor harvest, high tax imposts, and the egregious rates of interest moneylenders exacted to transform Armenian peasants into serfs under Kurdish and Turkish landowners. While migrant workers sought unskilled labor in cities to improve their finances in the village, others settled in larger towns and cities to swell the Armenian urban artisan middle class in Tabriz, Tiflis, Erzincan, Sivas, Kayseri and others, regionally, as well as Constantinople, Kafa, Lvov, Astrakhan, and Agra, etc., further afield. As a result of their new role in the economy, their values revolved more around capital than land, allowing them greater flexibility and mobility, attaining a new worldview and aesthetic, which, in turn, would find some reflection in the homeland.

In consequence, granted the degree of hemisphere-wide motion and transition that inevitably impinged on Armenia, too, arguably the best framework within which to discuss these developments would be one incorporating all nodes in the above networks within the overarching structure of an Armenian oikoumene, the contours of which were dynamically changing. This sort of synthetic approach would allow us to track movements, trace trajectories and gain meaningful insights into the complex processes that engender them. However, as indicated above, the period has not generally elicited sufficient interest to become the subject of a separate monograph, and, perceived as a chapter within a larger enterprise, has inevitably been accommodated to the overall design valorizing developments on the Plateau and severing diasporic issues from the main narrative on a center v. periphery model. Thus, volume four of the Soviet Armenian "History of the Armenian People" (*Hay*

---

[22] The phenomenal seafaring skill of Polynesian sailors across vast expanses of the Pacific Ocean with only rudimentary technology is a major exception to the above remark.

[23] Soil erosion and aridization arose from climate change and cooling, which precipitated the demise of agriculture, leading Armenians to settle in greater numbers in towns and Kurds to turn to animal husbandry to make a living. The "Little Ice Age" of the Northern Hemisphere that produced these effects began around 1300 with a recorded peak in 1550-1650.

Zołovrdi Patmut'yun) employs this as a foundational principle, dividing coverage into two main sections relating to a) homeland and b) diaspora. It further distinguishes political history from socio-economic and subdivides the latter between the Ottoman and Safavid spheres. Thereafter, the diaspora is presented in chapters broadly discussing events from the fourteenth to the eighteenth centuries, geographically segmented into the Near East, New Julfa and India, the Ukraine and Eastern Europe, and Russia. The achievements of Armenian communities in Western Europe and elsewhere are conspicuous by their absence.[24] Bournoutian follows a similar approach, presenting the sweep of Armenian history from the fifteenth to the nineteenth centuries, in a series of parallel chapters dealing with the Ottoman Empire, Iran, the Indian Subcontinent, and the Russian Empire, together with a joint study of Eastern and Western Europe. Meanwhile, the Dédéyan and Hovannisian volumes pair a chapter on sixteenth century events on the Plateau with one handling diaspora communities.[25]

In the absence of the rationale of an Armenian state as the organizing principle in this period, this nationalistically informed model has the unfortunate consequence of isolating the two spheres of activity, rather than integrating them, and of privileging one at the expense of the other. In turn, it underlines the need for maintaining a critical stance toward primary sources so as not inadvertently to perpetuate their rhetoric while drawing on them to gain an understanding of underlying events. A case in point is Karapet Bałiseci's lament of 1513 in immediate response to the devastation resulting from early Ottoman campaigns that portrays dispersion as inherently vulnerable and negative on the basis of Jewish precedents.[26] As more recent scholarship has demonstrated, diaspora is a phenomenon of enormous diversity that demands much more attentive interdisciplinary analysis.[27] In the dearth of contemporary sources embodying a broader, longer-term view of Armenians' sociopolitical and economic situation, it is incumbent on historians to proceed with appropriate caution.

Fiction as a Reflection of the Zeitgeist

So far, the discussion has focused largely on histories and historically related works in Armenian, mainly the product of the monastic clergy and inscribing its perspectives and predilections, although we must acknowledge that the latter represents one voice among many. What of the role of other literary genres, including in other languages, and translations as sources for this period, especially as they pertain to illuminating aspects of Armenian identity? In this connection, I would contend that, since fiction appeals to the imagination in visualizing the self in dialogue with the social environment presented there,

---

[24] Presumably this omission was politically inspired.

[25] Dédéyan, 341-76 and 377-409; Hovannisian, 1-50 and 51-79.

[26] On this poet, see Xač'atryan, *Hay mijnadaryan patmakan ołber (žd-žē dd.)*, 92-108; S. P. Cowe, "Print Capital, Corporate Identity, and the Democratization of Discourse in Early Modern Armenian History," *Le Muséon* 126 (2013): 340-45; Thomas, 582-87.

[27] On this general topic, see R. Cohen, *Global Diasporas: An Introduction* (Seattle: University of Washington Press, 1997).

it possesses rich possibilities both to mirror the status quo, as well as to plant the seeds for its transformation through its evocation of different ideals. If the chronicle's value lies in conveying accurate details about discrete events in specific locations, fiction opens up considerations of *longue durée* and insights germane to the *Zeitgeist* of a whole era. In order to provide this sort of illumination, it must be studied against the backdrop of the sociopolitical and economic ambience out of which it emerged. However, once again, this program of interface strives against the current norms of Armenian historiography where literature and the arts tend to be segregated from historical reality to be addressed in a hermetically sealed aesthetic realm.[28] Moreover, issues of translation automatically pose questions of how porous ethnoreligious boundaries can be, or put otherwise, how many converging levels of identification from the local to more distant environments can simultaneously coexist in a given society. Similarly, it is important to recall that, as noted previously, though written Armenian literature in the sixteenth century was still largely created and preserved by the clergy, there was a great degree of exchange between the written and oral traditions, and the growth of illuminated versions of certain works exposed them to a much larger 'readership'. Hence, a rigid distinction between clerical and lay taste in the cultural sphere is difficult to sustain.[29] At the same time, as the period is one of transition, one should observe the extent to which it serves as a watershed from one social and cultural aesthetic to another, particularly in the transference of patronage from the aristocracy to the middle class and the supersession of Persian preeminence by models of diverse literary provenance.

The Aristocratic Ethos of the Alexander Romance and the Near East

A case in point is the Alexander Romance. Already a legend in his lifetime, Alexander of Macedon evoked a series of literary treatments that gained enormous popularity. The Greek tale attributed to his court historian Callisthenes crossed into a number of languages by Late Antiquity, the Armenian version, produced at the end of the fifth century, possibly being the initiative of lay scholars, but thereafter transmitted in monastic scriptoria. Apart from its hero, the work's fascination rested on its combination of fact and fiction, travels to distant locations, and account of marvels. Consequently, it easily traversed religious boundaries to enter Islamic culture, in Persian and Ottoman versions, together with traditions in Arabic, Jewish texts, and Central Asian circles (e.g. the thirteenth century Mongolian version). Its appeal to the medieval aristocratic ethos inaugurated a series of further developments. In parallel with other traditions, the Armenian version spawned an

---

[28] See C. P. Ałayan et al, *Hay žołovrdi patmut'yun* [History of the Armenian People], vol. 4 (Yerevan: Armenian Academy of Sciences, 1972), 437-626; Dédéyan, *Histoire des Arméniens*, 369-72; Hovannisian, vol 1, 293-325; G. A. Bournoutian, 198-200.

[29] On the first collection of Armenian secular poetry from the seventeenth century, see S. P. Cowe, "Models for the Interpretation of Medieval Armenian Poetry," in *New Approaches to Middle Armenian Language and Literature*, ed. J.J.S. Weitenberg (Leiden: Rodopi, 1995), 40-41.

abbreviated medieval retelling,[30] and its most opulent manuscript copies were embellished with miniature illustrations. One might also argue the Hellenic prose romance entered a second, 'orientalized' phase of its Armenian transmission during the fourteenth century when tradents transformed it into the mixed media format typical of tales of Near Eastern origin, the original narrative becoming punctuated by images of its most emotionally evocative scenes, which were further interpreted by sets of verses. Begun in the 1330s, that lavish artistic treatment attained its zenith and its final flowering in our period, under the exquisite elegance and delicacy of Grigoris Ałtʻamarcʻi (c. 1480-1544), his pupil Zakʻaria Gnunecʻi (c.1500-c.1576), and the scribe and illuminator Yovasapʻ Sebastacʻi (c.1510-c.1564).[31] That they could accomplish this task patently indicates not all areas of the Armenian Plateau were devastated by military incursions, in the first half of the century.

A scion of the Sefedenian branch of the princely and royal house of Arcruni, Grigoris represents one of the few Armenian aristocratic families to cling precariously to their privileges through ecclesiastical sees, in this case the Catholicate of Ałtʻamar, which survived until 1895. Most of the new noble families that emerged in the eleventh and twelfth centuries to join the representatives of the older houses saw their estates and authority shrink over the fourteenth and the fifteenth centuries and presumably sold their properties and joined the exodus to the city. Consequently, within the wider trajectory from the late medieval to the early modern, the periodicity of the Armenian reception of the Alexander Romance well illustrates the transition in agency and patronage of culture from the nobility.

Medieval New Persian Genres and Armenian Literature

As the Armenian intellectual elite of the Late Antique period appropriated the structure of the Classical Greek educational system, along with its more diffused Christian culture in token of its identification with the wider norms of the eastern Mediterranean of that time, so we observe its participation in the Persianate cultural continuum of the medieval age, together with representatives of other Christian communities of the region such as the Georgian and East Syrian.[32] The exquisite burgeoning of New Persian poetry in the genres of lyric, epic, and romance established it as the consummate medium for poetic composition from the Near East to Inner Asia. This process was facilitated by the relocation of highly educated and administratively skilled Persians dislodged by the Khwarezmian incursions

---

[30] On this, see H. Simonyan, "Grigoris Ałtʻamarcʻin ibrev kafanerə 'Patmutʻyun Ałekʻsandri Makedonacʻwoy vepi kafaneri helinak" [Catholicos Grigoris Ałtʻamarcʻi's Kafa Verses in the History of the Great Emperor Alexander of Macedon], *Lraber* (1968), 446-89.

[31] On Grigoris' oeuvre, see Y. Kʻiwrtean, "Grigoris katʻołikos Ałtʻamarcʻii kafanerə 'Patmutʻiwn meci ašxarhakalin Ałekʻsandru Makedonacʻwoy' mēj" [Catholicos Grigoris Ałtʻamarcʻi's Kafa Verses in the History of the Great Emperor Alexander of Macedon], *Handēs Amsōreay* 81 (1967): 423-44; Simonyan, "Grigoris Ałtʻamarcʻin," 85-93. Three of his exemplars of the romance survive. They were executed in 1525, 1526, and 1536.

[32] On king Teimuraz I's compositions in this sphere, see D. Rayfield, *The Literature of Georgia: A History* (London: Garnett Press, 2010), 114-18.

and subsequent Mongol invasion, who assumed court positions among the Seljuqs, Ilkhans, and smaller dynasties in the region.[33] Consequently, Persian tropes like the love motif of the Rose and the Nightingale marked one of the new directions Armenian lyric elaborated from the thirteenth to the sixteenth century. Here, too arguably its finest exponent is Grigoris Ałtʻamarcʻi, who also finessed Persian and Armenian macaronic verse of great subtlety and refinement.[34]

In premodern Armenian society the genre of epic evolved in an oral matrix. The medieval composition, *Daredevils of Sasun*, transmitted by word of mouth into the twentieth century was in gestation around the time Firdausi created his literate masterpiece, *Shāh-nāma*, on the basis of earlier oral traditions.[35] As the latter circulated along the routes outlined above, Armenians created an oral version of part of its plotline entitled *Ṙostam Zal*. One of its distinctive elements is the fusion of characters and episodes derived from the indigenous Armenian epic.[36]

Arguably, however, the apogee of the Persian poetic genius lies in the third genre of romance, a genre that, though widely reflected in Georgia, previous scholarly consensus had asserted was not represented in Armenian society.[37] Though Boyce surmised that the Armenian Church exercised too powerful a moral control to permit such diversions,[38] we owe a popular late thirteenth century humorous version, somewhat satirical of the genre's high style and serious conventions, to the churchman and savant Yovhannēs Erznkacʻi.[39]

Persianate Romance in Armenian Redaction

However, I would suggest the most representative example of the genre is the *Patmutʻiwn Farman Mankan* (the Story of the Youth Farman), which probably also derives from the thirteenth century. It constitutes the first extensive poem in Armenian on the subject of secular love, whose popularity is vouched for by the scholar Yovhan Orotnecʻi of the second half of the fourteenth century, who cites the tale in a discourse analysis based on Aristotle's *Analytics* on the diverse types of syllogism. There he states,

---

[33] On this process, see C. Hillenbrand, "Rāvandī, the Seljuk court at Konya and the Persianisation of Anatolian Cities," in *Les Seldjoukides d' Anatolie*, ed. G. Leiser (Paris: Editions Hêrodotos, 2005), 157-69.

[34] See B. Čʻugaszyan, "Grigoris Ałtʻamarcʻu tałeri parskeren hatvacneri vercanumə," [The Decipherment of the Persian Portions in Grigoris Ałtʻamarcʻis Tał Poems], *Patmabanasirakan Handes* (1960): 201-22.

[35] See A. Yeghiazaryan, *Daredevils of Sasun: Poetics of an Epic* (Costa Mesa, CA: Mazda Publishers), 11-24.

[36] See B. Xalatʻeancʻ, *Irani herosnerě hay žołovrdi mēǰ* [The Iranian Heroes among the Armenian People] (Paris: Banaser Press, 1901), 45-79.

[37] On romance in medieval Georgia, see D. Rayfield, *The Literature of Georgia: A History*, 3rd ed. (London: Garnett Press, 2010), 75-95.

[38] For these views, see M. Boyce, "The Parthian Gōsān and the Iranian Minstrel tradition," *Journal of the Royal Asiatic Society* (1947): 15.

[39] For an English translation, see Cowe, "The Politics of Poetics: Islamic Influence on Armenian Verse," in *Redefining Christian Identity: Cultural Interaction in the Middle East since the rise of Islam*, ed. J. J. Van Ginkel et al. (Leuven: Peeters, 2005), 400-403.

> Now the second [category] is completely false, as books of mythology concerning Aramazd [Zeus] and Hermes [and] similarly nowadays the Copper City and Šeran Šah and others like them, for they are completely false [i.e. fiction].[40]

Typical of Persian romance, it is set in the pre-Islamic period, several of its features paralleling those of *Vis o Rāmin*, which is located more specifically in the Parthian era. Both narratives hinge on love between the scions of two royal families set at the geographical extremes of the Iranian world, presented as Assyria and Khorasan in the first and Media and Merv in the second, together with the trials and impediments to their love, and the lovers' final blissful union. Several of the tropes employed in the plot are also held in common: disguise, exchange of letters, temporary loss of one of the lovers and the other's lament on this, escape to a place of refuge, and the hero's succession by his son. Other motifs, such as the dream is reminiscent of *Khusrau and Shirin*, while the close camaraderie between the hero and his male companion P'ayip'ař resembles that between Avtandil and Tariel in Rustaveli's *The Knight in the Panther's Skin*. As the latter must continually console his friend as they search for his beloved, so the former must react to setbacks, strategize and execute operations to free his master from prison and unite him with his love, stave off disgruntled pahlivans, and prepare to meet the enemy forces. The fact that the love motif is treated directly in the absence of any mystical overtones seems to reinforce the comparatively early date of its core. Similarly, Davis suggests that, since a "happy ending" constitutes the norm in Greek romances but is featured in relatively few Persian works, this may be the result of interchange between the two traditions during the Parthian period.[41] At the same time, the plot differs from romance elements in the *Shāh-nāma*, in that the lovers' union is endogamous rather than exogamous and is the center around which the plot pivots rather than being orchestrated as the prelude to the birth of a hero.

Despite the ancient *mise-en-scène*, and the fact that the entire ambience and onomasticon is Persian, and that the work presumably derives from an Iranian matrix, this anonymous adaptation reflects many facets of the Armenian Middle Ages. This begins with the verse medium. Whereas Persian romances are composed in the *mathnawī* meter, featuring rhyming couplets, though the Armenian reflects the Persian eleven syllable line, it is couched in four-line stanzas marked by stanzaic rhyme, the most widespread form of Armenian poetry outside the most elevated. Also notable is the lack of any testing of the heroine's chastity, a quite common feature of the Persian genre, as in the episode of the sculptor Farhad in *Khusrau and Shirin*, which may therefore have formed part of the Iranian archetype. If so, this may have been altered to accommodate Armenian social norms.

In centrifugal Armenian society the image of royal etiquette the romance portrays would

---

[40] A. Minasyan, *Surb Hovhan Orotnec'i Meknołakan-imastasirakan čařer* [St. Hovhan Orotnec'i, Exegetical-Philosophical Homilies] (Church Fathers 4) (Ējmiacin: Mother See Press, 2009), 160. In the romance, Šah-i-šeran is Farman's son.

[41] D. Davis, *Panthea's Children: Hellenistic Novels and Medieval Persian Romances* (New York: Bibliotheca Persica Press, 2002), 1-7.

easily have been appropriated by nobles, and, in an era of mirrors for princes, the values held up for emulation are readable in the hero's characterization. These include leadership skills to promote good government through the acquisition of sound judgment to strategize and discernment to select counsellors to be effective experts in different affairs of state. They are spelled out most obviously in the scene where Farman picks four trusty attendants with different skill sets to act as his royal retinue on his journey to the east. Moreover, those qualities are inculcated through a more formal educational process. While in earlier times exposure to military training and horsemanship would have occupied most attention, here they emerge as the final phase of the prince's education, when he is already aged twenty. Most of his studies from age seven onwards have been spent with a royal tutor who guided his formation in such varied disciplines as medicine and divination. Moreover, not only is he surrounded by books in the palace but decides to take some with him on his journey to continue his cultivation of practical wisdom. That composite image is especially suggestive of the Hetʻumid house of Cilicia in the second half of the thirteenth century, since Levon II took great pains over his son Hetʻum's education under his secretary Vahram Rabuni, and the latter's homonymous relative Hetʻum of Kořikos reveals his fascination with the latest medical developments in copying an important miscellany on the subject in addition to his more well-known literary endeavors penning a recovery treatise for pope Clement V.[42]

Similarly, in contrast to earlier heroic treatments, the emphasis in Farman's duel with the champion Pʻolati Hndi, is on skill and ability rather than violence and brute strength, while extricating his master from prison and other impediments is mainly due to Pʻayipʻaŕ's resourcefulness rather than miracle or supernatural intervention. Moreover, despite her absence from the title, the heroine Šahišapnur plays a key role in the central part of the plot, initiating the relationship with Farman through her letter and visiting the youth in his quarters that night on his positive response.

Armenian aristocrats might well visualize themselves and their social values within the setting of this romance through the prism of its aesthetic distance from the everyday. However, by the sixteenth century, the bond between fact and fiction had been severed by the social and demographic transformation of that environment. No longer was castle or palace culture at the center of Armenian communal life, but, while Armenian peasants' rich vein of folklore continued in villages on the land, the expanded presence of the Armenian middle class in towns and cities created a demographic that demanded a new aesthetic from a different background.

Armenian Society and the Interaction of Politics, Economics, and Art

As literature and other cultural media cannot be compartmentalized from sociopolitical and economic developments, but rather reflect those in powerful ways, in this next section I wish to illustrate the varied interconnection between those spheres as literary works document new sociopolitical realities. The first sketch exploring this interplay relates to the

---

[42] S. Vardanian, *Histoire de la médecine en Arménie de l'antiquité à nos jours* (Paris: L'Européene d'Éditions, 1999), 97-104.

significant changes in identity that emerge in the late fifteenth and sixteenth centuries as Ottoman-Safavid socioreligious diversification evolves. Shah Isma'il's perception that the Turkomens of northwest Iran could be united by the Twelver branch of the Shi'ite creed to form the core of a new state proved unexpectedly effective, permitting him to expand the territory under his control to embrace regions of Southern Caucasia and eastern Anatolia with a large Sunni population. In turn, Sultan Selim had himself proclaimed Caliph and therefore defender of the sanctity of Sunni traditions and hence ordered the massacre of Shi'ite communities in the eastern territories he had recently captured. Those diverse steps inaugurated a state of fairly steady warfare throughout the sixteenth century, until a stable border was agreed upon by the Treaty of Zuhab, in 1639. This prolonged process instituted the latest phase of the Armenian Plateau's division by a fixed and frequently militarized boundary that has largely been maintained up to the present.[43] As some contemporary martyrologies affirm, this development, intensified by ideological animosities, interrupted generations of peaceful cross-border traffic, causing difficulties for both religious and secular travelers alike.

A martyrology of the bishops Yovhannēs and Sargis and the monk Dawit' composed in 1506 by the priest Yovhannēs treats the regular visit of a group of nuncios (*nuirak*) of the Armenian catholicos in Ējmiacin, in the Persian sphere, distributing newly consecrated holy oil (*miwron*) for use in baptism to various Armenian communities in the Ottoman Empire. One of them who had set out from Constantinople, ahead of the others, was stopped in Amasia and charged with espionage for the Safivids in league with European powers. The proceedings continued in the capital before various paşas and, finally, Bayezid II himself sentenced them to death. The charge of spying reveals a great deal about the timeframe in which they were caught up and recalls the precedent of a dual East-West offensive that had been investigated during the first Ottoman-Venetian War (1463-79) with Uzun Hasan of the Akkoyunlu.[44]

A martyrology by Yovasap' Sebastac'i, in 1536, deals with a similar situation. The youth Kawkčay assisted his father in his business by overseeing the transportation of commodities. His return home after an assignment was celebrated by loud festivities that so annoyed Muslim neighbors that they fabricated charges against him of having defamed the sultan and claimed the country belonged to the qizilbash. The ascription of pro-Safavid proclivities and insults against Sultan Suleyman seem to reflect continuing tensions deriving from the second Ottoman-Safavid War (1532-35) vented against a ethnoreligious minority maintaining too visible a profile and suspected of collusion with the enemy on the awareness of a large Armenian population on the other side of the border and a family enriching itself on long-

---

[43] For a map locating both sites in their sixteenth century context, see R. Hewsen, *Armenia A Historical Atlas* (Chicago and London: University of Chicago Press, 2001), 148.

[44] See Manandean and Ač'aŕean (eds.), *Hayoc' nor vkanerə (1155-1843)*, 340-45 for the text and Thomas, *Christian-Muslim Relations*, 572-75, for a brief study of its significance.

distance trade.⁴⁵ The role of Armenian clergy, especially Armenian Catholic, and merchants as intermediaries in such negotiations between the European powers and Iran, becomes important only later in the seventeenth and early eighteenth centuries.⁴⁶

The second scenario I would like to sketch relates to the intertwining of interests between the Armenian hierarchy and the papacy in the aftermath of the Ottoman-Safavid division of the Armenian Plateau, in the first decades of the century, and the contemporaneous eruption of the Protestant Reformation. The immediate initiative taken in closed conclaves at Ējmiacin (1547) and Sivas (1562) to promote the reestablishment of an Armenian kingdom with Western assistance, presumably against the background of the anti-Ottoman Holy League of 1538, involved only the traditional religious and secular elite. However, the request for the Venetian doge to facilitate contacts with the Holy See clearly builds on a long history of mercantile and financial contribution by the local Armenian community to the Republic's economic success. Similarly, Step'anos Salmastec'i's discussions with Sigismund II of Poland, in 1551, relied on the excellent relations the Armenian communities of Lvov, Kamieniec-Podolski, Zamość and others enjoyed with the crown, filling an important niche in the socioeconomic fabric of such international entrepots, as Gellner indicates.⁴⁷ At this point they were also absorbing into their internally autonomous quarters some of their fellow countrymen from the Plateau, as they approached the height of their prosperity, before countrywide movements toward religious conformity undermined their internal cohesion. The initiative failed to gain traction in part because the Armenian Plateau was now bestraddled by two powers, one of which, Iran, was increasingly viewed as a potential ally in containing Ottoman aspirations by opening a second eastern front.

At the same time, the Counter Reformation sought to reinforce the universality of the Roman Church in the aftermath of the Protestant Reformation by extensive mission and encouraging hierarchs of eastern churches to unite with Rome. The willingness of a succession of catholicoi of Ējmiacin and Sis to accept papal supremacy was obviously an important factor in Rome's diplomacy with Catholic monarchs. Moreover, as a series of Armenians was sent to Venice, after Yakob Mełapart's inauguration of Armenian printing in 1512, to acquire type and arrange the publication of further volumes, some of them were exploited to advance Catholic propaganda among their coreligionists. Yovhannēs Terznc'i is an excellent example. Together with his son, he arranged the second printing of the Psalter in Venice in 1587, which was important not only for liturgical purposes but as a basic primer in reading

---

⁴⁵ See Manandean and Ač'aṙean, 373-75 and V. P. Gevorgyan, *Hovasap' Sebastac'i banastełcutyunner* [Yovasap' Sebastac'i Poems] (Erevan: Armenian Academy of Sciences, 1964), 184, for the text, and Thomas, 644-49, for a brief study of its significance.

⁴⁶ On these activities, see R.H. Kévorkian, "Le négoce international des Arméniens au XVIIᵉ siècle," in *Arménie entre Orient et Occident*, ed. R. H. Kévorkian (Paris: Bibliothèque National de France, 1996), 142-51.

⁴⁷ E. Gellner, *Nations and Nationalism* (Ithaka: Cornell University Press, 1983) 58.

in widespread use in Armenian church schools.⁴⁸ However, his previous two publications were produced in Rome, at the behest of pope Gregory XIII, who was particularly zealous on Armenian outreach. The first was a translation of the pontiff's introduction to the new calendar that bears his name, which he had instituted in 1582 and was gradually adopted by Armenians in Europe over the next century to facilitate their dealings with Catholic states. The second was the translation of a catechism bearing the imprint of the Council of Trent patently intended to further the Catholic cause among Armenians.⁴⁹ Among the reforms approved at that council was the creation of new monastic orders like the Jesuits, who took advantage of Spanish and Portuguese navigational expertise to prosecute their mission in Asia and the New World. Disembarking at Goa and given accord in the Mughal capital, like the pioneer António de Andrade, they would join caravans to reach further into the interior of the continent where we learn they would periodically engage the services of Armenians familiar with the Inner Asian trade routes to guide them to Tibet.

The third sketch illustrates the interwoven tapestry of Armenian communities around the Black Sea and Eastern Europe, its texture continually changing in response to market conditions compounded by political and religious factors. This was possible because the merchant and artisan communities in question were not tied by bonds of land or loyalty to any one location and possessed marketable skills, which therefore favored movement and exchange. Whereas, once established, a core contingent tended to remain in a community, much of the population would fluctuate over time according to patterns that repay further study. One important component in this equation was Ottoman expansion, which led to a temporary decline in the Armenian population of the Crimea after 1475 and in that of Bulgaria in the last decades of the century. Similarly, in Poland we observe a remarkable growth over the sixteenth century followed by a decline in the following resulting from various factors, one of which was state and church pressure for the Armenian community to adopt Catholicism. One of the best illustrations of this arc is the city of Zamość. Founded in 1580, it experienced rapid growth in the early years of the following century during which an opulent central thoroughfare, styled Ormiańska Street, was constructed for an Armenian merchant community, many of whom thereafter relocated to Wallachia.

The operation of the above network is vividly documented by the martyrdom of the three nuncios of 1506 already addressed. Having paid their respects to the Armenian bishop in Constantinople, they first sailed to Kafa (Theodosia) on the Crimea and then southwest to Kara Boğdan (Turkish term for Moldavia) where presumably they disembarked at the port of Akkerman to travel to Suceava (contemporary capital of Moldavia) and then on to Lvov in Poland. From there, they retraced their steps to the port again from where they took ship back to the Ottoman capital. Such pastoral visits, some two generations later, would have followed a different routing. At the time of the visit in question, Suceava was a thriving independent capital, however its prosperity deteriorated under the rapid process

---

⁴⁸ On this, see N. Oskanyan et al., *Hay girkĕ 1512-1800 t'vakannerin* [The Armenian Book 1512-1800] (Yerevan: Myasnikyan State Library, 1988), 14-15.

⁴⁹ See Oskanyan, 12-13.

of Ottomanization exacerbated by a severe economic crisis. These facts are substantiated by the city's Armenian religious architecture, most of which dates to the first half of the sixteenth century. Thereafter, the center of gravity moved to Transylvania.

The fourth sketch relates to Armenian involvement in trade across the northern hemisphere. Armenian participation in international commerce can be traced at least to the regional economic upswing of the ninth century and experienced a major boost under the Pax Mongolica that supported the expansion of Armenian operations overland and, to a certain degree, via sea to China, as well as in the Black Sea and the Mediterranean, in association with the Italian maritime states. Armenians are documented as trading in Moscow in the fifteenth century and created a permanent community there and at Astrakhan, in the next. Those then benefitted from the consolidation of the Russian state to establish river and overland connections to Archangelsk with continuation by sea to various western European ports in the seventeenth century. Similarly, from tentative beginnings in the sixteenth century along parts of the east and west coast of India, Armenian investment in India led to the consolidation of earlier trading posts into more permanent communities thanks to the religious toleration extended by the Mughal Empire (c.1526-1757), a milestone in which was the foundation of the first Armenian church, in the capital Agra, in 1562.

This radical redrawing of the political and economic map of the hemisphere is matched in the Near East by the formation of the two further Islamicate empires of the era, the Ottoman (1453-1923) and the Safavid (1502-1736), we have already discussed. Whereas, so far, our attention has been drawn to religious and political aspects of their interface, this perspective must be complemented with the economic. The Ottomans gained Aleppo from Mamluk control in 1516, but the early phase of hostilities halted cross-border commerce in Persian silk, until it was designated an entrepot, in 1530. Since commerce was of mutual benefit, trade routes seem to have been respected by both sides during the wars and the region thrived so that by mid-century the city surpassed Damascus as an international hub. Commodities, both from Iran and Inner Asia, were mainly brought by Armenian merchants from the region of Naxčawan, Julfa, and Agulis and traded with representatives of a range of European states who protected their business investments by establishing consulates there—Venice (1548), France (1562), England (1583), and the Netherlands (1613). However, as Aleppo's economy depended on that of Iran, when the latter experienced a downturn at the end of the seventeenth century and a full-scale collapse over the Afghan Revolt of 1722, Aleppo was sent into a downspin until the mid-nineteenth century.

Nevertheless, during the era of the city's growth, the existing Armenian community expanded to become a major religious and cultural center. It is against that background that we are to interpret the rewriting of a major part of the Armenian redaction of the romance *Paris and Vienne*, produced by Yovhannēs Terznc'i, in 1587.[50] The earlier Western European versions portrayed the disgruntled lover Paris embarking on a Crusade after being rebuffed by his beloved's father in his suit for her hand. The inherent attraction this episode held

---

[50] On this, see L. Ter Petrosian, *Ancient Armenian Translations* (New York: St. Vartan Press, 1992), 41-2 and 108-09. Note that the English translation on p. 43 incorrectly states that the work was published in 1587.

for those readers was that of a journey into the unknown, filled with the mystique of the exotic marked by divergent mores and customs. Obviously, such an account would have provided a different aesthetic experience for Armenian readers very familiar with the terrain. Consequently, one facet of the work's indigenization is their inscription in the narrative as the hero sails to Cilicia and continues overland through the Armenian Plateau and Georgia to Iran, before descending from the old Safavid capital of Tabriz via Mesopotamia to Diyarbakir, at that time a strategic Ottoman border post,[51] and then on to Aleppo. The second part of his itinerary thus retraces the route currently plied by the Armenian silk merchants discussed above.

Subsequently, the protagonist embarks on a pilgrimage to Jerusalem and, en route, encounters a European Catholic cleric in Beirut. This would naturally align well with the original work's setting in the Crusades, but at the same time would evoke for contemporary readers the current context of the Roman mission in the Near East discussed above and Armenians' interaction with them there. This meeting marks the romance's turning point for the youth in apprising him of intelligence on his beloved's father Delphin. As a trusted aristocrat of the French crown, he had been assigned a secret mission to engage in espionage in the king's service, in the course of which he had been apprehended and incarcerated. It is interesting that, although such activities are not unknown in Crusading times, Terzncʻi identifies the destination of Delphin's mission as being Tʻurkestan, yet another contemporization dependent on Sultan Selim's campaigns of the 1510s extending Ottoman control to Greater Syria and Egypt. This, in turn, conjures images of the current activities of French diplomats and spies in the region and their varied contacts with the local Armenian merchant population.[52] Meanwhile, the hero (in disguise) is able to extricate Delphin from his predicament on his agreement to grant his daughter's hand to his rescuer, whose identity, at that point, he does not know.

Significantly, as the redactor informs us in his concluding colophon, the ambience in which he chose to fulfill his task was Marseilles, the hub of the French Atlantic fleet, which already contained the matrix of an Armenian community that was to gain greatly in importance in the following century as the port became the center for trade with India.

Similarly, the end of fifteenth century witnessed the creation of the nucleus of an Armenian community in Amsterdam engaging in the pearl and diamond trade by the second half of the sixteeth century. It also underwent major growth during the seventeenth as Holland became one of the preeminent maritime and economic forces of the age.

In this way, as the romance suggests, in the sixteenth century Armenian merchants were in mercantile contact with the English, French, and Dutch companies in Aleppo within a larger Mediterranean context where Venice still had a major role to play. Moreover, in

---

[51] The redactor had lived for some time in the city until his wife's death, after which he left for Venice with his son in the 1570s.

[52] Terzncʻi knew of what he wrote in this episode, having spent some time in a Roman jail himself, as he informs readers in the colophon to his publication of the Psalter in 1587. For details, see Oskanyan, *Hay girkĕ 1512-1800 tʻvakannerin*, 15.

that same century they had already begun to construct communities in Marseilles and Amsterdam and were likewise establishing permanent settlements on the Indian seaboard and other parts of East Asia. Over the next two centuries, they would form deeper relations with those same Atlantic powers in both the European and Asian spheres and, therefore, played their part in facilitating the emergence of the the inchoate world system.

The West European Romance Paris and Vienne in Armenian Redaction

The popular romance, *Paris and Vienne*, referenced above probably originated in Catalan before 1364, thereafter passing into Provençale, being rendered into French by Pierre de la Cépède in 1432, and subsequently published in Italian, Spanish, and Latin. A late medieval best seller that went through several printed editions over the fifteenth and sixteenth century, the tale is set in the thirteenth century and presents the fairly stereotypical plotline of an idyllic love between the two young aristocrats thwarted by the girl's father. This provokes the suitor to embark on a Crusade and experience various adventures in disguise in the course of which he rescues his beloved's father, as already mentioned, after eliciting his promise to grant him his daughter's hand. With the father now reconciled to the match, the wedding goes ahead smoothly. Subsequently, the couple bear several children and live to a ripe old age.

Granted the range of western romances available at that time, a series of qualities commend this particular example for rendition into Armenian. Firstly, even in its French prose version, it is significantly shorter than many others and would have been even more condensed in the Italian verse intermediary from which the Armenian redaction probably derives. Similarly, as Cooper argues, it lacks the episodic plot typical of the genre, and, being free of marvels and wonders, possesses more verisimilitude and is consonantly one of the first romances to depict the characters' feelings in a more naturalistic manner.[53]

As a bestseller, it is likely that Terznc'i accessed the text through the contemporary medium of print.[54] A married priest from the area of Bitlis, an important commercial center at this time, he chose to redact the work at a bustling port not in a monastic library.[55] In so doing, he demonstrated noteworthy agency in undertaking the task of his own volition rather than acceding to the behest of an ecclesiastical superior, as would have been the previous norm, and therefore marks an additional facet of the piece's modern ambience. Moreover, as his vocation brought him closer to the ethos of the lay community, with a family of his own, he would possess more empathy for the theme than one of the celibate hierarchy. In

---

[53] H. Cooper, "Malory and the Early Prose Romances," in *A Companion to Romance from Classical to Contemporary*, ed. C.J. Saunders (Oxford: Blackwell Publishing, 2004), 118.

[54] For Terznc'i's biography, see Ł. Ališan, *Hay-Venet* [Armenian Venice] (Venice: St. Lazar's Press, 1896), 210-15, and Melik'-Ōhanjanyan, 1966, 7-36.

[55] The term the redactor applies to his activity in the final colophon is *tramadreal* ("having transposed") rather than merely "translated." See K.A. Melik'-Ōhanjanyan (ed.), *Patmut'iwn P'arēzi ew Vennayi, Patmut'iwn Venetik K'ałak'in* [The Story of P'arēz ew Vennay, The Story of the City of Venice] (Yerevan: Armenian Academy of Sciences, 1966), 233.

this, as we have observed, he was representative of one of the broader tendencies of the age that witnessed a greater democratization of writing inevitably incorporating a wider spectrum of subject matter in a simpler style and a register combining the classical idiom with features of the vernacular.[56] Emanating from the city and smaller towns, often with the taste of the merchant and artisan middle class in mind, it more readily embraced more secular pursuits. Indeed, the poetic treatment of humorous topics like Minas T'oxat'ec'i's mock-heroic 'Panegyric of Harisa' (traditional winter delicacy) of 1563 or his compatriot Step'anos' 'Complaint on Lice and Flies' of 1614 prepare the way for Martiros Łrimec'i's more trenchant satire later in the seventeenth century.[57]

It is clear that the redactor operated with his countrymen's proclivities in mind. In this connection, it is noticeable that early Armenian printed books usually run less than a hundred pages so as not to tax readers' patience.[58] This, he replicated to create a lighter, simpler, fast-paced treatment. Moreover, the tendency to reproduce the piece with illuminations may suggest that, in many cases, one person might recite the text aloud, while the others present followed the action through the images. The redactor achieved his goal by reducing details of the French aristocratic ambience still imbedded in his Italian text to render the work more accessible to his mainly middle-class readership. Thus, he reduces the role of the pope in organizing a Crusade and abbreviates much of the circumstantial detail pertaining to a joust the hero attends near the beginning of the tale. This had been part of the court life of the Armenian kingdom of Cilicia but was now an outmoded spectacle that readers would be unable to relate to. Hence, his excision of the incident parallels Cervantes' roughly contemporary critique of such practices in his masterpiece, *Don Quixote*. Gone are the long dialogues, counsels, and lovers' speeches, as well as commonplace tropes, such as Vienne's birth to her previously childless parents after a wait of seven years, thereby advancing the plot by an emphasis on action.

Another means of rendering the work more contemporary was the elimination of certain indications locking the plot into the thirteenth century so that it acquired more of an indeterminate character with which readers could identify more readily. Similarly, whereas earlier forms of the romance only mention Paris learning "Moorish," i.e. Arabic, before setting out in the original context of a Crusade and contacts along the eastern littoral of the Mediterranean, Terznc'i adds Persian in acknowledgement of that idiom's wide subsequent dissemination, which we have already addressed. Moreover, the importance the theme accords to language acquisition would resonate with readers who would be required to learn a range of languages themselves to travel smoothly in Western Europe, as Terznc'i himself commanded at least Latin and Italian.

Unlike the redactor of the earlier romance *The Story of the Youth Farman*, Terznc'i harmonizes some of the realia to Armenian practice to further readers' imaginative

---

[56] Cowe, 339-44.

[57] On the poet, see A.A. Martirosyan, *Martiros Łrimec'i: usumnasirut'yun ev bnagrer* [Martiros Łrimec'i: Study and Texts] (Yerevan: Armenian Academy of Sciences, 1958).

[58] Cowe, 322-25

identification. Thus, the names of some of the secondary characters appear in their Armenian form as Yovhan (John) and Yakob (James), while the narrative medium retains the traditional structure of eleven-syllable verse with stanzaic rhyme. We have already discussed some of the commercial overtones of the hero's journey to Aleppo, however his disembarking in Cilicia and travel to Greater Armenia, where he visits Ani, which he qualifies as the *mec amroc'* ("the great fortress"),[59] powerfully resonates with initiatives to reestablish Armenian statehood with western assistance in mid-century, which we have already discussed. As illustrated by voices like that of Karapet Bałišec'i, pressuring the elite to take action, the ruins of Ani, capital of the Bagratid kingdom (884-1045), exerted a potent stimulus on the imagination of the intelligentsia from at least the time of Nersēs Šnorhali (mid-twelfth century) to seek to reactivate that symbol of Armenian statehood.[60] Similarly, Terznc'i contextualizes Paris' journeys within the concept of łariput'iwn, a state of wandering in foreign parts, exposed to potential dangers and hardships one must face alone, far from home and family and a support network.[61] Widely reflected in various literatures of the Near East in this period, it is particularly applied to the situation of economic migrants in the Armenian tradition, whose numbers were increasing in the general exodus to the cities. Moreover, in his peregrinations the hero gains contemporaneity in an era typified by mobility that was now being inscribed in autobiographical treatments of long distance pilgrimage, as in Martiros Erznkac'i's visit to Santiago de Compostela in the 1490s and Simēon of Zamość's travelogue, embracing many sites along the northern and eastern shores of the Mediterranean.[62]

Cooper classifies the French romance as a precursor of the novel, its publication placing it in the hands of an upwardly mobile middle class.[63] Certainly, the Armenian version affords the hero even more scope for character development in his protracted travels since he journeys on his own with no comrade to release him from difficulties. As he lives in this foreign environment for several years, he constantly has to face the issue of trusting people with his identity and eking out his finances to survive. In contrast to the much more conventional portrayal of the earlier hero Farman, his later counterpart undergoes valuable psychological development and thereby becomes much more individualized as a character, triggering empathy with merchant readers.

---

[59] Melik'-Ōhanǰanyan, 200.

[60] Cowe, 335-40.

[61] Significantly, the redactor's companion conjures that mindset in describing his situation in Marseilles as being "continually worn down in this foreign realm" (yōtarut'iwns mišt mašec'ay), see Melik'-Ōhanǰanyan, 233.

[62] See Hakobyan, *Manr žamanakagrut'yunner XIII-XVIII dd.*, 97-110, and Bournoutian, 25-313.

[63] Cooper, 107, 117.

Significantly, despite the variety of European adaptations, the Armenian is the only Near Eastern version[64] indicating its importance in the new paradigm of East meets West in which Armenian penetration of European markets parallels its drawing on the region's cultural values.[65] The encounter heralds a new phase of Armenian translations encompassing a larger variety of secular themes over the next two and a half centuries, obviously in response to lay interest and increasingly through lay mediation from Latin, as well as from vernacular sources in English, French, Italian, Polish,[66] and Spanish, a number of which were also abbreviated in Armenian.[67] Consonant with this trajectory is the impact of Neoclassism on Armenian aesthetics of the eighteenth century where, once again, it represents the only Near Eastern literary tradition to resonate with that largely European artistic movement.

As we have noted, the sixteenth century, for the Armenian Plateau as for the Armenian oikoumene more broadly defined, is one of immense transition and change, transforming older patterns of thought and action together with cultural paradigms in keeping with wider global trends. The redaction of *The Story of the Youth Farman* well illustrates the Armenian reflex of New Persian cultural influences in circulation across the region from the east over the thirteenth to sixteenth centuries. This also constitutes a major watershed in the history of Persian poetry, the classical era of which ends in the fifteenth century, to be followed by the very different Indian style of the sixteenth. Moreover, as the aristocratic ambience that engendered it passes, we observe the timeliness of the redaction of *Paris and Vienne*, emerging from a western background as reflected in the *mise-en-scène* that in its new form represents more of a middle-class ethos with merchant proclivities.

Aşık Romance in the Armenian Tradition

As the Armenian redaction of *Paris and Vienne* illustrates the transition from aristocratic to middle class mores, the transfer of authorship from the monastic to married clergy, and the supersession of a Persian aesthetic by one from Western Europe, concomitantly a third

---

[64] The only other version produced outside Western Europe is Vintzenzos Kornaros' seventeenth century Greek adaptation entitled *Erotokritos*, an influential poem of the Cretan Renaissance based on an Italian form available on the island from when it belonged to the Venetian thalassocracy.

[65] With Terznc'i's adaptation of the European romance for his readers, one might parallel the reception process in other media, such as Yovhannēs Mrk'uz's adaptations of Dutch models for his fresco cycle of St. Gregory the Illuminator's conversation of Armenia in the All Savior cathedral in New Julfa, in which he indigenizes the iconography by touches like making the figures somewhat more squat and including the twin peaks of Mt. Ararat in the background. This then became emblematic of Armenian handling of woodblock prints in eighteenth century printing.

[66] One of the first of these is the *Hayeli varuc'* [Mirror of Mores], translated by Simēon Lehac'i in the seventeenth century from the Polish version of a Latin original providing moral instruction through vignettes from social situations. See K'. Ter-Davt'yan, *Hayeli varuc'* [Mirror of Mores] (Yerevan: Armenian Academy of Sciences, 1994)..

[67] A good example of this trend is Yovhannēs Avdalean's rendering of Defoe's *Robinson Crusoe* into Classical Armenian in Calcutta.

type of romance evolved in a Turkic environment, also important for the study of history, which is distinguished by the primary role of lay bards, implying a further democratization of literacy. With origins in the sixteenth century, in Anatolia and Southern Caucasia, in a swathe that embraced the Armenian Plateau and straddled the Ottoman-Safavid political divide, this genre would continue to grow over the next three centuries. While continuing to utilize some of the medieval Persianate subject matter, it developed in an urban, rather than a courtly environment. Not only did it present the outlook and values of the middle class, but it was usual for exponents to learn a trade to supplement their income from performance. A frequent milieu for the latter was the coffeehouse, itself a recent innovation after the beverage was declared licit for Muslims in the sixteenth century.[68] Moreover, the varied inflections of Turkic formed the compositional norm as being the lingua franca of the region. However, in keeping with the above, its primary focus was love rather than the heroic exploits and endless battle scenes that had filled the plot of premodern Turkish epics like the Battal-nāme and Danişmend-nāme. In contrast, all the heroes of these new narratives were lovers, as were the raconteur-poet-performers in their collective persona as *aşık* (Armenian *ašuł*), a term derived from the Arabic term for love (*aşk*). While the precise prose form to be recounted orally would be formulated in extempore fashion, on the basis of established contours, the songs, intercalated at points of heightened pathos, would be written in the poet's *daftar* (notebook). The repertoire spanned a wide variety of verse types, resulting in a very flexible medium for the modulations of romance that would leave a deep emotional impression on the urban audience.

Another facet of Armenian agency in this period is the participation of poet-performers in this new literary movement evolving on the Plateau. Moreover, since it was the norm to compose in Turkish, Armenian bards would often write all or most of their songs in that idiom, especially in the early period, as in the case of the first documented Armenian *ašuł* Nahapet Kučʻak, who is so designated on his tombstone of 1591. Of the nine poems that have come down to us under his name, two are Armenian, while the other seven are in Turkish.[69] Armenian employment of local languages in various capacities is paralleled, for example, by officials of the Armenian court at Lvov keeping their records in the Turkic language of Kipchak, which was also widely used in the Armenian community on the Crimea.[70] These texts would normally be penned in Armenian letters, giving rise to a phenomenon called Armeno-Turkish, which enjoyed wide circulation over the next three centuries. Moreover, in his travelogue, Simēon of Zamość notes that Armenian communities

---

[68] Association with the coffeehouse is reinforced by mention in some of the narratives on which see I. Başgöz, *Hikâye: Turkish Folk Romance as Performance Art* (Bloomington: Indiana University Press, 2008), 241-42.

[69] On this, see Ō. Eganyan, "Nahapet Kučʻaki hayatar tʻurkʻeren tałerě" [Nahapet Kučʻak's Armeno-Turkish Tał Poems], *Banber Matenadarani* 5 (1960), 466.

[70] On these, see V. Grigoryan, *Kamenecʻ-Podolsk kʻałakʻi haykakan datarani arjanagrutʻyunnerě, XVI d.* [Armenian Court Records from the City of Kamenets-Podolsk (16th c.)] (Yerevan: Armenian Academy of Sciences, 1963).

in several parts of Anatolia spoke Turkish rather than Armenian, observing a process that continued to spread over the next two centuries.[71] To meet their pastoral needs, the Armenian Church began publishing religious materials in the Armeno-Turkish idiom, while its use for disseminating literary works is testified by the activities of the prolific lay writer Eremia Č'ēlēpi K'ēōmiwrčean who, among other writings, translated the Armenian redaction of *Paris and Vienne* into this medium as a further indication of its popularity.[72] The trend for ašułs to transfer most of their narratives and poetry to Armenian occurred in the course of the nineteenth century under the increasing impact of modern nationalism.

Armenian Redaction of the Köroğlu Romance

Although, as noted, a number of the themes for aşık romance were of older origin, one of the most popular subjects, that of Köroğlu (Armenian Kyoṙołli), emerged out of the contemporary maelstrom of the Jelali Revolts of the end of the sixteenth century.[73] Although the initial revolt had occurred in 1519, led by an Alevi preacher Celâl, who gave his name to the movement as a whole, the more serious outbreaks ensued over the years c.1590-1610 and 1622-59. The former was precipitated by troop disbandment after a lull on both the western and eastern fronts that compelled the discharged *sekbans* (irregular musketeers) and *sipahis* (calavry rewarded with land grants) to resort to banditry to compensate for their loss of wages. Together with semi-nomadic Turkmens and Kurds, the Jelali bands fomented revolt against currency depreciation, official bribery and corruption, and ever-higher taxes levied to finance the Habsburg and Safavid campaigns. Greater reliance on Janissaries led to the *sekbans'* demobilization, while the impoverished peasantry became uprooted from the land in hopes of finding more lucrative employment in the towns and cities, where the massive influx only resulted in more elevated levels of unemployment there, too, preparing the groundwork for a further escalation of the insurrection.[74]

This process, which indelibly transformed the demography of Anatolia and Southern Caucasia as a whole, as well as its Armenian community, exerted a lasting existential impact on the popular psyche that expressed itself in a fictionalized dramatization of the era, with

---

[71] Language became an ideological issue for the first time in Armenian society with the Mkhitarist Congregation in the eighteenth century, whose founder, Mxit'ar Sebastac'i, produced grammars of Classical and modern vernacular Armenian for this constituency to combat that tendency.

[72] His Armeno-Turkish version was published in Constantinople in 1871. See H. Stepanian, *Hayatař t'urk'eren grk'eri matenagitut'yun 1727-1968* [Bibliography of Armeno-Turkish Books 1727-1968] (Yerevan: Armenian Academy of Sciences, 1985), 84, no. 374. For the characteristics of the published text of the Armeno-Turkish version, see Stepanian, *Littérature arménienne en langue turque écrites en caractère arménien* (sic) (Yerevan: L'Académie des Sciences d'Arménie Institut d'Orientalisme, 2012), 65-7.

[73] For a recent reprint of episodes of the Armenian Köroğlu romance composed by Ašuł Jamali and Łazaros Ałayan, together with translations of Turkish, Azerbaijani, and Kurdish versions, see L. K'ot'anjyan (ed.), *Kyoṙołli* (Eastern Romances, 4) (Yerevan: Sayat'-Nova Cultural Union, 2007), 17-515.

[74] See W. J. Griswold, *The Great Anatolian Rebellion, 1000-1020/1591 1611* (Islamkundliche Untersuchungen, 83) (Berlin: K. Schwartz Verlag, 1983).

all its military and moral confrontations encapsulated in a real historical figure. Köroğlu, a Jelali leader of the sixteenth century from the area of Bolu in the vilayet of Kastamonu, was in fact a soldier-bard, as the historian Aṙakʻel Dawriẑecʻi records, who in actuality composed various *xał* songs before himself becoming the subject of a complete aşık cycle.[75] The latter reflects the cut and thrust between those in revolt and local authorities in the implacable hostility between the swashbuckling, larger-than-life bandit with his base of operations at Šamlibil and inseparable companion, his horse Ḻṙatʻ, against his adversary Boli Beg. The latter embodies the opportunistic perspective of his class in illegally raising taxes on the peasantry, while his opponent emerges as their protector who thwarts the beg's measures by guerrilla warfare and eases the locals' hardships by raiding merchant caravans whose transit fees would benefit the authorities. At its core, the narrative thus focuses on essential moral issues. However, this serious preoccupation is tempered by an infusion of humor and love interest developed in a series of amorous adventures. Moreover, as a fictionalization of contemporary events, the narrative medium is relatively realist, dispensing with the traditional repertoire of dreams, fairies, and the supernatural intervention of Khiḍr as a modernist enterprise.

The Köroğlu narrative attained an unprecedented diffusion in thirteen languages not only throughout the Near East but also in Inner Asia,[76] where a distinct variant developed.[77] Incipient comparative studies have revealed the degree to which the hero becomes indigenized in each of those, as, for example, in being identified with Iranian epic motifs in the Tajik version.[78] In Armenian, the protagonist, once more, represents a foreign hero of romance, whose reception in this case involves not only his molding to exemplify ethnic traits and mores but his complete identification as an Armenian. In this form the narrative has been transmitted from the seventeenth century up to the present.[79]

---

[75] For Aṙakʻel's Armenian text, see L.A. Xanlaryan, *Aṙakʻel Dawriẑecʻi Girkʻ patmutʻeancʻ* [Aṙakʻel Dawriẑecʻi: Book of Histories] (Yerevan: Armenian Academy of Sciences, 1990), 108, and, for an English translation, Bournoutian, *The History of Vardapet Aṙakʻel of Tabriz*, vol. 1 (Costa Mesa, CA: Mazda Publishers, 2005), 70.

[76] For an early transcription of an eastern Turkic version of the romance, see A. Chodźko, *Specimens of the Popular Poetry of Persia* (London: Oriental Translation Fund of Great Britain and Ireland, 1842), 3-344.

[77] On this, see Kʻotʻanjyan, *Kyoṙołli*, 5-6.

[78] See J. Wilks, "The Persianization of Köroğlu: Banditry and Royalty in Three Versions of the Köroğlu Destan," *Asian Folklore Studies* 60 (2001): 305-18.

[79] For the discussion of the more recent ašuł tradition, see S. P. Cowe, "The Art of Actuality: Contemporary Dastan of an Armenian Ašuł," *Edebiyat* NS 4 (1993): 117-29. The author also videoed four renditions of the romance in fieldwork conducted in October 2017.

Transition to the Novel

As Cooper highlights the role of the prose romance in Western Europe as the precursor of the novel,[80] in the Near East, the aşık's *hikâye* performed a similar function. A long prose narrative with a simple plot and style, focusing on roughly contemporary social and political realities and composed in the current vernacular, featuring credible characters neither fully idealized nor caricatured, and disseminated mainly in an urban milieu, the genre incorporated a number of the novel's primary characteristics. These were not lost on Xačʻatur Abovean, as he mused on what format to employ in presenting Armenian readers with a modern story of the archetypical human issues of love, death, and struggle so amply treated in foreign novels, but lacking in the sort of Armenian literature he had been exposed to. Appreciating the possibilities of the ašuł's narrative, particularly its immediate emotional impact on the audience, he sketches a typical scene in the introduction to the novel *Vērkʻ Hayastani* (Wounds of Armenia) of 1840,

> Many times, I have seen in gatherings and in the street how they gazed wonderstruck at a blind ašuł, paying heed, collecting money for him, the saliva running down their chin. When would they break bread at a party or wedding without a musician? His words were Turkish, and many could not understand at all, but the listener's, the spectator's soul would soar to paradise and return. I pondered, I pondered. One day I said to myself, "Come on, ditch your grammar, rhetoric, and logic.[81] Put them aside and you become an ašuł as well. Let what will be, be!"[82]

True to his word, Abovean adopted the basic ašuł narrative structure and the model of heroes like Köroğlu as a freedom fighter in molding his protagonist Ałasi, who is also an ašuł and sings to the accompaniment of the tar̄, at various points in the plot.[83] His oeuvre builds on Armenian contacts with the Russian court, dating back to the late seventeenth century, to recreate Armenian statehoood under Russian suzerainty, in hopes for better conditions for his people from a Christian power, devoid of the poll tax and other facets of subordinate status dhimmis endured under Islam. Begun with the premodern goal of reestablishing a dynastic kingdom, the campaign was gradually transformed under the impact of Romantic nationalism, of which Abovean became one of the first Armenian exponents. Exposed to the German model of this movement during his studies at Dorpat University, he became imbued with the principles of ethnography, celebrating the values and lifestyle of the people on the land as the true bearers of the national character through their folk customs, pithy sayings, etc., which he extolled in his folkloric scene painting. Indeed,

---

[80] Cooper, "Malory and the Early Prose Romances," 118-19.

[81] As a schoolteacher, Abovean's classes were given over to grammar, rhetoric, and logic, the basis of the Late Antique *trivium*, which was still formed the rudiments of the school curriculum in his time.

[82] The passage is taken from the preface to the novel.

[83] Note the parallel influence of the ašuł Jivani of Alexandrapol on the composer Armen Tigranyan in creating the score for the opera *Anuš* in 1912.

the novelist incorporates Gellner's portrayal of the classic Romantic ideologue's transition from an ecclesiastical calling to pursuing a secular career in education or journalism.[84] This, in turn, reflects the transformation of Durkheim's axiom, that through the practice of religion, the state worships itself indirectly, as under nationalism that self-worship becomes completely unabashed, and patriotism comes to demand of the individual the same self-sacrifice previously reserved for martyrdom to defend a religious creed.[85] Abovian's hero thus embodies the new patriotic ideal not only in his national rather than local affiliation and the righteousness of his allegiance to an army of liberation rather than engagement in small-scale skirmishes with a band of irregulars, but in his ultimate willingness to give his life to capture the Erevan fortress, the last bastion of resistance, putting his fellow countrymen's future before his personal fulfilment with his beloved. In this way the author presented his readers with a new image of heroism to emulate in constructing their communal identity, thereby entering the mainstream of nationalist discourse, perpetuating its message through the persuasive dissemination of potent emotionally appealing cultural icons and symbols.[86]

In this connection, it is significant that the first novel composed by an Armenian in the Ottoman sphere, Yovsēpʻ Vardanean's *Akabi Hikayesi* (Akabi's Story) of 1851,[87] was written in Ameno-Turkish and may also owe something to the tradition of ašuł romance as well as the Turkish translation of European novels.[88] Treating the struggles between two factions within the Armenian Catholic community of Constantinople and broader tensions between Armenian Catholics and the Armenian Apostolic Church, the work contributes to current debates relating to the role of religious confessionality within Armenian society in the broader context of the modern nationalist cause.[89] The transference of these internal religious divisions to the Ottoman administrative sphere further intensified the search for secular alternatives to anchor and unify the nationalist project that expressed itself powerfully in contemporary efforts to democratize the political process and enact internal social and fiscal reform. The effervescence of Romantic Nationalism in the next decade both in the literary and other cultural domains in turn articulated a much clearer concern with territorial issues and foregrounded cohesion and cooperation as watchwords for the future. Vardanean's work is therefore an important contribution to the nineteenth century development of a modern prose idiom as a vehicle for the novel and the dissemination of a nationalist discourse in the Ottoman-Armenian milieu.

---

[84] Gellner, 60.

[85] Ibid., 56.

[86] B. Anderson, *Imagined Communities: Reflections on the Origin and Spread of Nationalism* (London, New York: Verso, 1991), 80-2.

[87] For the publication of this work in current Turkish transcription, see A. Tietze, *Akabi hikayesi: ilk Türkçe roman, 1851* [The Story of Akabi: First Novel in Turkish, 1851] (Istanbul: Eren Yayıncılık ve Kitapcılık, 1991).

[88] On the translation of European novels into Armeno-Turkish, see Stepanian, *Hayatař tʻurkʻeren grkʻeri matenagitutʻyun 1727-1968*, 68-77.

[89] H. Barsoumian, "The Eastern Question and the Tanzimat Era," in *The Armenian People from Ancient to Modern Times*, R.G. Hovannisian, vol. 2 (New York: St. Martin's Press, 1997), 185-91.

Conclusions

Clearly the series of Ottoman-Safavid wars that punctuate the 16th century caused great disruption to the tenor of life on the Armenian Plateau, which is reflected in the decline in the typical monastic activities of manuscript copying and history writing. Nevertheless, I would contend that there are many more sources relating to the activities of the Armenian oikoumene, embracing both the Plateau and the different diasporan nodes than were perhaps previously recognized. Moreover, the Armenian community was engaged in such a varied range of pursuits over that century that, if it was legitimate for Cardinal Baronius to apply the test of manuscript production as a metric of intellectual vitality in Western Europe in the periods bookending the Carolingian Renaissance, it is difficult to justify that standard in our case, granted the diversification of Armenian society, its increased urbanization, the greater spread of literacy and numeracy among the middle class, and the democratization of writing to include married priests, deacons, and *dpirs* (clerks), who would be educated and function in a city or town environment without direct contact with monastic establishments.

Similarly, granted the wider scale of Armenian enterprises, the traditional outlook of monastic historians with its limited perspective requires supplementation by a series of other genres, including fiction, which, as we have seen, was largely consumed by lay audiences in an oral setting, as well as later in printed form.[90] Moreover, Yovhannēs Terznc'i's initiative in editing and redacting *Paris and Vienne* began a process of modern translation from other languages in which lay scholars subsequently participated, introducing Armenian readers to works that increasingly embodied secular themes and values. Similarly, contemporary lay ašułs, such as Nahapet Kuč'ak, composed narratives and poetry treating themes of love and heroism that anticipated first Eremia Č'ēlēpi K'ēōmiwrčean (1637-95) and, later, novelists like Xač'atur Abovean. Consequently, building on the relatively recent awareness that the romance genre is represented in Armenian society, we should grant it a more central place in discussions of literature and aesthetics.

Those achievements in the literary plane parallel others in the sociopolitical and commercial realms as a ringing affirmation of Armenian agency in an important era of transition. Far from Armenians entering into hibernation or living an inward-looking ghetto-like existence, we observe their activities on a global scale responding to local, regional, and hemispheric trends, participating in a series of longer-term enterprises that only achieved fuller momentum and, therefore, visibility over the following century.[91]

Consequently, the interrelation between different facets of Armenian society, the interface between developments on the Plateau and in different diaspora centers, and the coherence of all those communities within wider networks of exchange, demand that we investigate those complex processes in a more synthetic, interdisciplinary manner. As

---

[90] For the 1709 publication of the Armenian version of the *Copper City* and a series of other Near Eastern tales, see Oskanyan et al, *Hay girkĕ 1512-1800 t'vakannerin*, 177-78.

[91] For the transition from the sixteenth to the seventeenth centuries in Julfan commerce, see S.D. Aslanian, *From the Indian Ocean to the Mediterranean* (Berkeley and Los Angeles: University of California Press, 2011), 23-85.

Dadoyan has reminded us of the embeddedness of Armenia and Armenians within the Arab and Islamic world during the medieval period,[92] we would do well to heed the utility of Subrahmanyam's insights in approaching the sixteenth century, and the Armenian Early Modern period more broadly, from the perspective of connected history.[93]

---

[92] See, for example, S.B. Dadoyan, *The Armenians in the Medieval Islamic World, Paradigms of Interaction Seventh to Fourteenth Centuries*, 3 vols (New Brunswick and London: Transaction Publishers, 2011-2014).

[93] See, for example, Subrahmanyam, *Explorations in Connected History*. Note that Sebouh D. Aslanian was first to argue for the utility of Subrahmanyam's connected histories approach to Armenian Studies in general and especially to the early modern period in Armenian history in public fora and conferences. See further Sebouh Aslanian, "From Autonomous History to Interactive Histories: World History's Challenge to Armenian Studies," in *An Armenian Mediterranean: Words and Worlds in Motion,* edited by Kathryn Babayan and Michael Pifer (London: Palgrave Macmillan, 2018): 81-125.

SOME REMARKS ON THE IDENTITY AND HISTORICAL ROLE OF ARTSAKHI MELIKS IN THE SEVENTEENTH AND EIGHTEENTH CENTURIES CE

Roman Smbatyan

The ideas of independence and statehood have been alive within the Armenian socio-political and cultural realm throughout the centuries, in the late medieval and early modern periods. The Artsakhi Meliks were at the forefront of this notion, and multi-dimensional analysis of their identity reveals not only their role in holding Artsakh as a formidable stronghold against foreign and local incursions but indicates their sense of historical role in liberating the Armenian lands. This paper tries to analyze the identity of Artsakhi Meliks on three levels of perceptions: self-perception, Armenian Diaspora perception, and foreign perception. While historians have studied the military, political and cultural role of Artsakhi Melikates extensively, very few general remarks have been made about how their identity was perceived in the above-mentioned dimensions. Furthermore, this paper tries to shed light on the identity of Artsakhi Meliks by focusing on Armenian, Russian and Persian primary sources, and archival materials, revealing the different perceptions of Meliks presented in these sources.

After the fall of the Armenian Kingdom of Cilicia, in 1375, and the rulership of Zakaryans, in 1350, Armenia once again lost its independence and fell under the rule of foreign empires. As sources state, in the late medieval and the early modern periods, when Armenia was divided between the Ottoman and the Safavid Empires, traces of small local Armenian autonomous rulerships were preserved in some mountainous regions of Western and Eastern Armenia, namely in Artsakh. Artsakh has been the easternmost province of Greater Armenia, today widely known as Karabakh and located in South Caucasus, between Armenia, Azerbaijan, and Iran. Ulubabyan, Walker, Chorbajian, and others propose that the word Artsakh derives from "Ardakh," "Urdekhe," "Atakhuni" mentioned in Urartian cuneiform inscriptions.[1] However, the final and definitive etymology of the name Artsakh is uncertain.[2]

According to Anania Shirakatsi, Artsakh was the tenth province of the kingdom of Greater Armenia until the first division of Armenian lands between the Sassanid and Roman Empires, in 387 A.D., when, as Toumanoff mentions, Artsakh and Otene (Utik) became a part of Caucasian Albania.[3] Later, under the suzerainty of Sassanid Persia, Armenian Arranshahiks ruled in Artsakh in rivalry with Iranian Mihranids, the rulers of the eastern parts of Arran.[4] Beginning in 821 A.D., when the Armenized Mihranid dynasty's rule on Arran ended, local Armenian commander Sahak Smbatyan, more famous under the Arabized name Sahl Ibn Smbat, founded the local kingdom of Khachen or Artsakh.[5] He was a descendant of the Armenian local dynasty of Arranshahiks and was also regarded as the prince of Armenia by contemporary Byzantine rulers. Arabic geographers mention the name Arran and Shirvan to describe historical Albania, while in Armenian sources the name Khachen, Artsakh, and other local names are used to describe the region. The gradual Turkification of some parts of Arran and Shirvan started at the beginning of the eleventh century, with massive inundation of Turkic and Turkmen tribes into the Caucasus. As Barthold states, the name Arran is rarely mentioned in contemporary historical sources, since the Mongol domination in the region, and the name Karabagh instead is used in Persian sources to describe the southern part of historical Arran.[6]

The name Karabagh itself is formed from turko-mongolic *kara, which means "black, big," and Persian *bagh, meaning "garden." The land of Artsakh is rich in dense mountainous forests which conditions the toponymy of the region. As to Meliks, the word "melik" is the Armenized version of Persian word "malek," which itself is formed from Arabic root word *mlk and means "king", "prince", "local ruler." According to "Karabagh-Name" by Mirza Jamal Jevanshir Gharebaghi, the title "melik" was given to the local rulers of Artsakh by Safavid Shahs, namely Shah Abbas, at the beginning of the seventeenth century, when the latter was satisfied with the invaluable support provided by Artsakhi rulers and Armenians overall during the war against the Ottomans. Moreover, the Armenian archbishop requested the Iranian Shah to liberate Armenians from the Ottoman yoke.[7] Shah Abbas, and later Nadir

---

[1] Bagrat Ulubabyan, *Artsakhi Patmutyuny Skzbits Minchev Mer Orery* (Yerevan: M. Varandyan, 1994), 12-13.

[2] For further discussion, see David M. Lang, *The Armenians: A People in Exile* (London: Unwin Hyman, 1988), x.; Movses Khorenatsi, *Hayots Patmutyun*, trans. Stepan S. Malkhasian (Yerevan: HaypetHrat, 1961).

[3] Cyril Toumanoff, "Introduction to Christian Caucasian History: The Formative Centuries (IVth-VIIIth)," *Traditio* 5 (1959): 75.

[4] H.S. Svazian, *Aghvanits Ashkharhi Patmutyun* (Yerevan: HH GAA "Gitutyun", 2006), 225-30.

[5] Bagrat Ulubabyan, *Khacheni Ishkhanutyuny X-XVI darerum* (Yerevan: Haykakan SSH GA, 1975), 64-73.

[6] Vasili V. Barthold, *Sochineniya, vol. III* (Moskva: Nauka, 1977), 335.

[7] Mirza Jamal J. Gharebaghi, *Tarikhe Gharebagh* (Tehran: Markaze Chap o Entesharat, 1994), 10-12.

Figure 1. Melik Egan's Castle in Togh village. Accessed from http://www.monuments.nkr.am

Shah, believed that only by restating the rights and positions of loyal local leadership, could they retain control over the Caucasus and elsewhere.[8] Thus, it was during the seventeenth century, and especially the first half of the eighteenth century, that the influence and power of the Artsakhi Meliks peaked. Not only they could preserve their power and lands in Artsakh and bordering territories, but the Khamsa Melikate of Artsakh was formed as a completely autonomous region under the direct rule of Nadir and his army commander with the privilege granted to Meliks to levy taxes, conduct justice, hold military forces, and protect the Armenian population.[9] But what was the identity and background that made the Meliks of Artsakh the most viable military and political force in war-torn, weakened, and divided Armenia?

After the erosion of the united Armenian statehood, following the invasions of the Seljuk Turks and the Mongols, the Artsakhi Meliks were the most independent of all analogous Armenian principalities and, as Robert Hewsen, mentions, "they saw themselves

---

[8] Roman Smbatyan, "Siasate Nader Shah Dar Jabejayie Aghvam va Houzeye Edarie Ghavghaz," *Farhange Mardom (Be yade Iraj Afshar)* (2011), 114-25.

[9] Artak Maghalyan, *Artsakhi Melikutyunnery ev Melikakan Tnery XVII-XIX dd* (Yerevan, 2007), 31-7.

as holding onto the last bastion of Armenian independence."[10] Some important evidence can be mentioned that describes the Meliks' identity on three basic levels: self-identity, identity perceived by the Armenian Diaspora, and identity perceived by the foreigners.

Meliks considered themselves as the holders of political and military power needed to protect, although on the very local and isolated level, the tradition of Armenian statehood. One of the most prominent Artsakhi Meliks of the eighteenth century, Melik Egan's inscription on the wall of the presence chamber of his house in the village Togh names him the khan and beglarbeg of Khamsa melikate. He not only compares himself with Georgian valis (kings), but also claims that he did not allow any Armenians to be captured and enslaved by foreigners. Moreover, on the gravestone of the same Melik, we find the following verses:

> This is the tomb of the brave prince,
> The great melik named Egan
> The son of the honorable clergy named Ghukas,
> He was loved by Nadir Shah and all the people,
> He owned the land of Aghvank with provinces
> He was greatly honored by the Persians as the melik of the Armenian land.[11]

These verses state that the prominent status of Melik Egan was considered not only a significant achievement at a local level but also on the broader Armenian context. As we can see here, the name Aghvank is used to describe Artsakh, which is consequently mentioned as part of Armenia. Thus, Melik Egan himself and his relatives regarded him as a greatly honored leader of Armenian land. From the foreign perspective, the author of the most important Persian source on Nadir Shah's period, Mohammad Kazem Marvi, mentions that Melik Yegan was the most powerful leader (*rīšsefīd*) among Armenians and had very close relations with the Iranian shah.[12] Furthermore, with direct support of Nadir Shah, five melikdoms of Artsakh could strengthen their position and, in the late 1730s, founded a new administrative organization called Moluke Khamsa, thus uniting the five Armenian melik authorities of Karabagh. Nadir assigned the most influential Armenian, Melik Yegan, as head of Moluke Khamsa. Yegan gained all the administrative rights of the *beglarbeg*, a local head of a region in the Iranian state who was in charge of gathering taxes. Moreover, Melik Yegan was in charge of conducting justice and punishing the guilty.[13]

Another contemporary of Melik Yegan was Varanda's Melik-Husein Melik-Shahnazaryan on whose tombstone we can find the following text:

---

[10] Robert H. Hewsen, "The Meliks of Eastern Armenia: A Preliminary Study," *Revue des Études Arméniennes* 10 (1972): 52-3.

[11] H. Papazyan, "Melik Yegani Yndunarani Mutqi Vimagir Ardzanagrutyuny," *LHG* 5 (1985): 78.

[12] M.K. Marvī, *Alamaraye Nāderī* (Tehran: Nashr-e Elmi, 1979), 410.

[13] Roman Smbatyan, "The Position of the Caucasus in the Policies of Iran of Nadir's Era," (PhD diss., University of Isfahan, 2010), 155.

> This is the tomb of Melik Husein,
> The son of Melik Shahnazar,
> Let us glorify his deeds
> And for his honor this inscription is written on his grave stone,
> He was the Melik of the land Varanda with thiry-five villages,
> His table was full of food and he gave alms to everybody,
> His appearance was handsome, he didn't pay tribute to any king,
> And he was the powerful stronghold of all the country,
> The pride of Armenian people that defeated the Ottomans.[14]

Thus, even in this period, defeating the Ottomans was supposed to be a national pride for the Artsakhi Meliks and the Armenians. Besides, most powerful Meliks were almost always identified as symbols of Armenian statehood and military power. It is not surprising that, while negotiating his plan of liberation of Armenia with the European leaders and the Russian Tsar as the representative of Armenian Meliks and archbishop, Israel Ori emphasized on the armed forces of meliks as the core military power capable to contribute to the liberation of the whole Armenia[15] The Artsakhi Meliks were also using diplomatic sources to propose their accomplishment by sending envoys to foreign states. Additionally, they had close ties with the leaders of the Armenian Diaspora in Russia, Iran, and India. From this viewpoint, noteworthy is the letter by well-known political thinker of Indian Armenian Diaspora, Shahamir Shahamiryan, addressed to the bishop Hovhannes of Gandzasar, which says, "Even with erosion of statehood, wealth and famous people, with the help of God you have not collapsed completely as Israelis, Egyptians, Assyrians or Greeks, but we still possess local autonomy".[16] No doubt, Indian Armenian thinker meant the Meliks of Artsakh as the last stronghold of Armenian statehood and the base for possible future Armenian independent state. As observed later, Shahamiryan acts as intermediary to reconcile Melik-Abov of Gyulistan and Melik-Mejlum of Jraberd against Ibrahim Khan of Karabagh.[17]

Even before Melik Egan, Artsakhi meliks were on the frontline of the Armenian liberation movement, the active phase of which started in the early 1720s, in Artsakh and Syunik. A significant fact that proves the magnetic force of Artsakh in organizing the liberation of Armenia is that, as a result of social, political, and religious pressure by the Safavid government, in 1717, Avan Yuzbashi moved to Varanda melikate of Artsakh from Lori and began to organize the liberation movement of Armenia. Avan Yuzbashi was one of the commanders of ten thousand Armenian military forces that headed to Cholak to join Peter's army in Northern Caucasus, which was supposed to liberate Armenia from

---

[14] Raffi, *Erkeri Zhoghovatsu, vol. 9* (Yerevan: Sovetakan Grogh, 1987), 52.

[15] *Hay Zhoghovrdi Patmutyun, vol. 4* (Yerevan: Haykakan SSH GA, 1972), 138-44.

[16] Maghalyan, 36.

[17] Ibid., 36-7.

Iranian rule.[18] Moreover, he was not only organizing the resistance in Artsakh but in 1724 together with his forces he moves to Syunik to help David Beg in the fight against local khans assigned by Safavid Shah. Overall, the analysis of Avan Yuzbashis scope of activity shows that, although being a local Melik, he was active everywhere where the question of liberation of Armenia was raised.

From the perspective of self-identity of Artsakhi Meliks, another important fact can be mentioned. During the liberation movement in Armenia, in the first half of the eighteenth century, Artsakhi Meliks and arhcbishop of Gandzasar who, as a rule, was a representative of Khachen Meliks' family of Hasan-Jalalians, had accepted a statute of twenty-three articles, the most important of which stated as follows:

Each melik should recognize the Jalalian meliks as leaders of the whole Artsakh and recognize the Church of Gandzasar both as apparent and secret place of meetings of Artsakhi leaders. Each commander must obey their melik and each melik is obliged to possess regular military force of 3000 people. Each melik must protect their territory and hold the solidarity and in case of threat inform other meliks. The traitors of Fatherland or religion will be punished according to the decision of the joint committee of clergy and meliks. No melik can establish relative ties with foreign princes or join them. And most important is the following article: Each Melik is obliged to subordinate the interests of their region to the interests of Fatherland, as well as self-interests to the interests of their region.[19]

The last article states, "Any of us who violates these rules or joins the enemies of Fatherland or our religion and disturbs the peace in our land and hounds Armenians against each other will be sentenced to death and other Meliks will take control of their land and property."[20]

As we can observe, this is a set of rules for a monarchial system in the state of war. It clearly states the Meliks as responsible for the future fate of their people. Meliks of Artsakh deeply believed in their unique historical mission as protectors of Armenian statehood and independence. It is noteworthy that, at this same period, Armenian Catholicos and meliks who were sending their diplomatic envoys to different countries to seek help in liberation movement, had very close ties with prominent Diasporic thinkers in India and elsewhere. However, on the practical level, Artsakhi Meliks relied solely on their own military forces and will of fighting and protecting their Armenian homeland and identity. The liberation movement of Armenia eventually reached its active phase in the 1720s, thanks to the meliks of Artsakh and Syunik.

The rebellion of Karabagh showed that, despite foreign rule for centuries, the idea of independence, statehood, and deliverance were still alive in the Armenian entity and were key factors of the identity of Artsakhi Meliks and other Armenian leaders both in their

---

[18] Ulubabyan, 141-42.

[19] S.R. Hasan-Jalalian, "Katoghikos Esayi Hasan-Jalaliani Qaghaqakan Gortsuneutyunn u Patmagrakan Zharangutyuny," *Herald of the Social Sciences* 3 (2011): 84-6, URL: http://lraber.asj-oa.am/6040/1/2011-3_(82).pdf

[20] Ibid.

homeland and Diaspora, which had a strong impact on later history of the Armenian people. Even with the fact that Artsakhi Meliks were eventually defeated by the Ottomans, after almost a decade of resistance, Armenian Meliks showed farsightedness and allied with Nadir shah and melik Egan of Dizak became the ruler (*beglerbeg*) of united Karabagh (Moluk-e Khamse).[21] Armenians, in all over the country and region, were inspired to contribute to the later development of the liberation movement and the fulfilment of centuries of hope to revive the independent Armenian state.

Non-Armenian sources also support the unique role of Artsakhi meliks as Armenian leaders and protagonists of Armenian independence. The commander of Russian army in the Caspian Region, General Dolgarukov, in his report of 11 May 1727, states, "It is beyond a human mind to understand how they can resist such a mighty enemy."[22] In the second half of the eighteenth century, another well-known Russian, Generalissimo Suvorov, wrote in his papers, "After Shah Abbas' rule of Great Armenia only autonomous Karabagh is left."[23] Moreover, in 1783, Russian statesman Prince Potyomkin, in a letter addressed to his cousin, General Pavel Potyomkin, wrote, "The Khan of Shushi Ibrahim should be deposed, as Karabagh hereafter will become an Armenian independent region only under the rule of Russian Empire."[24] As history showed later, in the nineteenth century, the strong tradition of Artsakhi statehood and military made a significant contribution to the liberation of Armenia from Iranian rule.

Thus, the identity of Artsakhi Meliks, analyzed on different levels of perceptions, both by Armenians in their homeland and the Diaspora, as well as by Russian statesmen and Persian historians and, most importantly, the Meliks themselves, shows that, in the early modern period, the Artsakhi Meliks were considered as one of the main protectors of the identity and tradition of the Armenian statehood. Not only they possessed military force and fought to preserve their autonomy in Artsakh, but they also were at the forefront of the liberation movement of Armenia from foreign rule.

---

[21] Roman Smbatyan, "Nadir's Religious Policy towards Armenians," in *Studies on Iran and The Caucasus* (In Honor of Garnik Asatrian), eds. Uwe Bläsing, Victoria Arakelova and Matthias Weinreich, with the Assistance of Khachik Gevorgian (Leiden: Brill, 2015), 135.

[22] Ulubabyan, 145.

[23] Archive of National Academy of USSR, ph. 99, op. 2. no. 13. 19-59, 65-73.

[24] Ioannisyan, 68-9.

MODERN

ARMENIANNESS REIMAGINED IN ATOM EGOYAN'S *ARARAT*

Myrna Douzjian

*The imagination is now central to all forms of agency, is itself
a social fact, and is the key component of the new global order.*
                                                                -Arjun Appadurai

How has the "Catastrophe" shaped the identities of Armenians in the diaspora?[1] Scholarship published during the last decade of the twentieth century produced remarkably consistent responses to this question. For example, in the field of sociology, Anny Bakalian concludes, "The Genocide and its subsequent denial by Turkish governments is [...] a symbol of collective Armenian identity. [...] It serves as a common denominator, an equalizer of all differences between Armenians: national, regional, religious, ideological, political, socioeconomic, generational, and so on."[2] In a similar vein, but in the field of

---

[1] Here, I use the word *Catastrophe* in keeping with Marc Nichanian's argument that the more common designation, "the Armenian Genocide," constitutes a misnomer, because, as a positivist designation, it cannot describe the inexplicable destruction brought upon Armenians in the Ottoman Empire during WWI. Nichanian has painstakingly shown that Catastrophe, the English equivalent of the Armenian word *Aghed,* is the most apt designation for this moment, because it was the word used by survivors and witnesses themselves; moreover, Catastrophe or *Aghed* appropriately suggests that the event escapes the grasp of language. For a delineation of the ontological differences between genocide and Catastrophe, see Marc Nichanian, "Catastrophic Mourning," in *Loss: The Politics of Mourning,* ed. David L. Eng and David Kazanjian (Los Angeles: University of California Press, 2003), 99-124; Marc Nichanian, *The Historiographic Perversion,* trans. Gil Anidjar (New York: Columbia University Press, 2009); David Kazanjian and Marc Nichanian, "Between Genocide and Catastrophe," in *Loss: The Politics of Mourning,* ed. David L. Eng and David Kazanjian (Los Angeles: University of California Press, 2003), 125-47.

Throughout this paper, I occasionally opt for the terms Genocide and Armenian Genocide to draw attention to the common contexts in which these terms are used.

[2] Anny Bakalian, *Armenian-Americans: From Being to Feeling Armenian* (New Brunswick: Transaction Publishers, 1994), 356.

literature, Lorne Shirinian writes, "The Armenian Genocide has become a collective symbol which reorganizes the discourse pertinent to Armenian collective identity. In other words, the Genocide explains what Armenians are like and what the world they inhabit is like; furthermore, it also presents a paradigm of how they act."[3] These strikingly similar assertions represent a dominant line of thought that emphasizes the past-oriented nature of Armenian diaspora communities and therefore makes a case for their homogeneity. Despite the dramatic changes that the geographic centers of diaspora culture have undergone since the turn of the century, the Armenian collective consciousness continues to be framed in these terms.[4]

Cultural production, driven by the insatiable demand for narratives about the Catastrophe, also tends to present a homogenous view of the diaspora and its relationship to history. In the specific case of cinema, one-dimensional approaches to the impact of genocidal events have reaffirmed the unity of the diasporic experience. For instance, *The River Ran Red,* a documentary based on Michael J. Hagopian's collection of over 400 survivor testimonies, concludes with a telling oversimplification, expressed by the narrator-director, "They had one story to tell the world. We are the survivors of a simple truth."[5] Such a conclusion eliminates the complexities in survivors' memories in the service of reproducing a factual grand narrative that proves the Genocide.[6] The consistency of this narrative then becomes the basis for expressions of collective existence: since all survivors share the same story, their progeny inherits the same relationship to the "truth." Maintaining this truth in response to the Turkish government's denialism has been a priority for Armenian diaspora communities, who have favored, if not downright fetishized, historiographical scholarship and cultural products that affirm the factuality of the Genocide. That is to say, sweeping

---

[3] Lorne Shirinian, *Writing Memory: The Search for Home in Armenian Diaspora Literature as Cultural Practice* (Ontario: Blue Heron Press, 2000), 41-2.

[4] In contrast to the discourse that equates diasporic identity with genocidal history, some scholars have articulated nuanced understandings of diasporic communities. Most notably, Khachig Tölölyan acknowledges the role of competing, and therefore heterogeneous, allegiances among intellectual and political elites in the disparate groups that make up the Armenian diaspora. See Khachig Tölölyan, "Elites and Institutions in the Armenian Transnation," *Diaspora* 9.1 (Spring 2000): 107-36.

Despite the availability of critical approaches that attend to the complexities of diaspora, one of the predominant narratives in North America conceives of the Genocide as the foundation of Armenian identity. For example, Rubina Peroomian cites the Genocide as one of the most important sources of Armenian identity in an article published in the March 26, 2005 Armenian issue of *Asbarez Daily Newspaper*. In a more recent interview with Ara Tadevosian, published online at *Media Max* on January 13, 2015, Jirair Libaridian discusses the importance of undoing the popular notion that institutions in the diaspora are centered on and unified by the politics of Genocide recognition.

[5] *The River Ran Red,* directed by Michael J. Hagopian (Thousand Oaks, CA: Armenian Film Foundation, 2009), DVD, 56:26-56:32.

[6] Nichanian has argued that the appropriation of testimony, as evidence for "pseudo-historical ends," constitutes a historiographic perversion. Nichanian, *The Historiographic Perversion*, 103.

documentary accounts of the Genocide reproduce themselves and reinforce sweeping statements about Armenian identity in a type of inextricable union.

Narratives that do not belong to historically grounded subgenres, which include biography, memoir, and documentary, have garnered less interest and attention. Even the major motion picture *The Promise,* released in April 2017, was marketed as a historical drama that gives viewers access to the truth.[7] The film and footage from its production were later incorporated as material in Joe Berlinger's documentary, *Intent to Destroy,* which also features archival footage, photographs, and expert interviews. The appropriation of a dramatic film as a means to assert facts illustrates the lack of interest in the conceptual possibilities that fiction might offer to representations of genocide.

This description of the general attitude toward the representation of the Catastrophe prefaces my reading of Atom Egoyan's *Ararat* (2002) because the privileging of what we might call "proof narratives" adversely affected the film's reception. In contrast to the enthusiasm with which Armenian audiences have embraced documentaries like Andrew Goldberg's *The Armenian Genocide,* Eric Friedler's *Aghet: A Genocide,* Michael J. Hagopian's *The Witness Trilogy,* and, more recently, David Lee George's *Architects of Denial,* Egoyan's drama received a lukewarm response at best.[8] A structurally complex film, *Ararat* presents two films, both with the title *Ararat*. Edward Saroyan (played by Charles Aznavour) directs the embedded film, which depicts—in the dramatic style typical of Hollywood historical epics—the Armenians' act of self-defense in Van in 1915; Egoyan's *Ararat* portrays the making of Saroyan's film and the stories of the lives connected with it. The *mise en abîme* highlights the films' diametrically opposed aesthetic visions: while Egoyan's *Ararat* creates an occasion for debating the ethics of representing the Catastrophe, Saroyan's *Ararat* overwhelms any critical engagement through its spectacularization of the Genocide.[9] Though the two films are easily distinguishable, the inclusion of multiple personal and familial stories and intertextual references complicates the film's structure and contributes to the indeterminacy of its meaning. As a result, Egoyan was often criticized for not representing the Genocide as a clear-cut historical fact. One common reaction to the film's indeterminacy was summed up

---

[7] For one representative example, see Daniel Halton's article, "A Promise Fulfilled" in the February 2017 issue of *AGBU Magazine*. The Armenian Genocide Committee, a coalition of Armenian cultural, political, religious, and professional organizations, promoted the film's much-anticipated release alongside its annual Genocide recognition efforts. Irrespective of its artistic merit, the film enjoys the community's unbounded enthusiasm precisely because of its embrace of the one true Genocide narrative.

[8] Of course, some critics recognized the merits of *Ararat* and produced thought-provoking research that dealt with the film's treatment of postmemory, trauma, reconciliation and representation. See, for example, Elke Heckner's "Screening the Armenian Genocide: Atom Egoyan's *Ararat* between Erasure and Suture," Marie-Aude Baronian's "The (Im)possible Representation" and "Image, Displacement, Prosthesis: Reflections on Making Visual Archives of the Armenian Genocide," Hamid Dabashi's "Vision Impossible: Atom Egoyan's *Ararat,*" Anahid Kassabian and David Kazanjian's "*Ararat*'s Hands," and Marc Nichanian's "Representation and Historicity."

[9] Jonathan Romney, *World Directors Series: Atom Egoyan* (London: British Film Institute, 2003), 182.

in a review published in *The New Yorker,* in which Anthony Lane writes, "Egoyan could, and should, have made an exemplary documentary on the subject of the slaughter, and asked Aznavour—an Armenian born in Paris—to talk us through it. If I were a Turkish official, I would not be too worried about this picture. Nothing so slippery can stir up indignation."[10] Many viewers and critics like Lane wanted an expository documentary. When instead they got a psychological portrayal of characters coming to terms with and reinventing history, disappointment followed.[11]

To complicate matters further, many interpretations of the film contradict one another, and even its harshest critics give different rationales for their respective critiques. Among these critiques, Asli Daldal's is one of the most troubling,

> By failing to pay attention to historical nuances, by relying heavily on appeals to "post-memory" and nostalgia, and by repeating stereotypical and outmoded clichés about Turks reminiscent of films like *Midnight Express,* Egoyan's film seems rather to be intent on "preserving" the denial that it is ostensibly condemning. The political identity of the Armenian diaspora is arguably strengthened and consolidated by Turkey's politics of denial: Turkey is the common enemy and the Other against whom the diaspora confirms its imagined homogeneity.[12]

Like the criticism expressed by the proponents of the documentary genre, this interpretation suggests that *Ararat* perpetuates the denial of the Genocide; however, it offers the opposite reason in support of this claim. These two perspectives—the one criticizing the film for not being straightforward and simple enough and the other criticizing it for its nationalist one-dimensionality—exemplify the polyphonic, and at times cacophonous, reactions that *Ararat* elicited.

What, then, can be made of the coexistence of these mutually exclusive interpretations? If the urge to control historical discourse did not predetermine critics' reading of the film, it certainly influenced their evaluations of it. These responses demonstrate the continued politicization of cultural and historical representation in our ever-globalizing world. *Ararat* became the target of various attacks precisely because it subverts the clear-cut binary oppositions (self/other; truth/lie; history/fiction) that have informed critics' interpretations. What these critics miss is that Egoyan's rejection of a linear narrative about the Catastrophe and its denial should be read as indispensable to his unique approach to Armenian cultural

---

[10] Anthony Lane, "Worlds Apart: *Far from Heaven* and *Ararat,*" *The New Yorker* 78.35 (Nov 18, 2002): 104-05.

[11] The diaspora echoed Lane's sentiments, both in the press and at public lectures. Many Armenian diasporans insisted that Egoyan should have produced Saroyan's *Ararat* with all the bells and whistles of Hollywood filmmaking. According to these ill-disposed critics, Egoyan's *Ararat* was too convoluted, lacked far-reaching public appeal, and did little, if anything, for recognition efforts. Anahid Kassabian and David Kazanjian, "*Ararat*'s Hands," *Armenian Review* 49, no. 1-4 (2004-2005): 139-40.

[12] Asli Daldal, "*Ararat and the Politics of 'Preserving' Denial,*" *Patterns of Prejudice* 41, no. 5 (2007): 409.

identity. The film's convoluted story shifts the responsibility of meaning-making from its plot to its characters. *Ararat* conveys its messages through the characters' attitudes toward the Catastrophe in order to suggest that individuals have the ultimate agency in constructing both history and identity.

The film articulates its position on identity through a series of scenes wherein characters approach an understanding of themselves in one of two ways: by relying on predetermined historical narratives or by engaging actively in producing their own narratives. In particular, three self-reflexive scenes, when placed in conversation with one another, reveal the stark contrast between these two approaches, allowing the film to interrogate dogmatic notions about identity. The first scene begins with an exchange between the art historian Ani (played by Arsinée Khanjian) and Rouben, the screenwriter of Edward Saroyan's epic film. It takes place right after Ani's stepdaughter Celia vandalizes Arshile Gorky's painting, *The Artist and His Mother,* in an attempt to punish Ani for her indifference to the accidental death of her father. Appalled by Celia's act, Ani explains its severity to Rouben, "That painting is a repository of our history. It's a sacred code that explains who we are, and how and why we got here." She then walks onto the rolling set of Saroyan's film, thereby hypocritically tearing into a cinematic scene just as Celia tore Gorky's work of art. The disruption she causes infuriates one of the actors, Marty, who plays the role of the American missionary doctor Clarence Ussher. Although Ani's entrance puts an immediate end to the filming, Marty remains in character, describing the dire situation he faces as the savior of Armenian victims; at the end of his speech, he demands that Ani explain her position. "Who the fuck are you?", he asks.[13] His question draws attention to the centrality of the complex task of defining the self in the film.

The confrontation between Marty and Ani frames identity as a performance. Marty relies on a premade script as the ultimate actor—he has embodied his extensively researched role so thoroughly that he assimilates Clarence Ussher's persona. His self-important performance of the white savior, both in and out of character, reveals his reliance on this narrative trope for his sense of self.[14] In other words, the story he enacts offers the material for the performance of his own identity. Though she is at odds with Marty in this scene, Ani performs her Armenian ethnic identity according to an equally scripted narrative. Just a moment before being faced with Marty's question, Ani confidently articulates her ideas about her identity. However, when faced with Marty's question, she is stunned to the point of speechlessness.[15] The juxtaposition of Ani's silence with her previous claims about her identity invites a reconsideration of the latter. Her statement about the sanctity of Gorky's painting and its relationship to her people indicates that she sees herself as part of a larger collective *explained by* the Catastrophe. Just as Marty relies on the film script to define his character, Ani relies on Gorky's familial and artistic history to inform her sense of the

---

[13] *Ararat,* 1:22:32.

[14] In addition to being utterly predictable, this trope in film typically draws attention away from the nuances in the victimized group's experiences.

[15] *Ararat,* 1:19:00–1:23:00.

Armenian people and her sense of self. Because Marty/Ussher sees himself as the protector of the Armenians and Ani sees herself as the protector of Armenian culture and history, they both strive to enshrine the scripts they hold dear. Marty and Ani's parallel overreliance on a ready-made understanding of the self calls into question the formulaic nature of their characters.

Occurring at the midpoint of the film, the interaction between Marty and Ani functions as a focal point that illuminates other scenes' treatment of the performance of identity. One such related exchange that deals with staging and set design begins when Ani sees the set of Saroyan's *Ararat* for the first time. She points out that Mt. Ararat, the backdrop for the set, would not actually be visible from Van. She expresses her disapproval of the mountain's placement, and during the ensuing conversation, Ani and Rouben discuss their opposing views on artistic freedom:

> ROUBEN: Well, [Saroyan and I] thought we could stretch things a bit. It's such an identifiable symbol, and given the moment in history we're trying to show...
> ANI: It's something you could justify.
> ROUBEN: Sure. Poetic license.
> ANI: Where do you get those?
> ROUBEN: What?
> ANI: Poetic licenses?
> ROUBEN: Wherever you can.
> *Pause. ROUBEN stares at ANI, slightly on edge.*
> ANI (smiling): So that's my job? To let you feel better about distorting things?[16]

This scene, critical of both Rouben and Ani's perspective, represents a telling example of the film's evenhanded treatment of its characters. Ani's refusal to entertain, for even a moment, the spectacular, "untrue" additions to Saroyan's film reveals her inflexibility yet again. At the same time, her objection to the geographic distortion in the film draws attention to its artificial additions meant to arouse predictable, emotive audience responses. In fact, during the premier of Saroyan's *Ararat,* even the filmmakers and the lead actor respond to the scenes depicted onscreen in the same way, crying and cringing in unison.[17] Thus, the disagreement between Rouben and Ani also offers a key to what each of their approaches misses. First, Rouben (and Saroyan) should have the authority to imagine, but not formulaically and not for crudely manipulative purposes. As Egoyan himself explains, Mt. Ararat is the "most fetishized" Armenian symbol.[18] Given the proliferation of the mountain in cultural products, its placement in the film functions like a Pavlovian sign rather than an artistic statement. Second, Ani's fixation on the truth causes her to uphold an oversimplified

---

[16] Atom Egoyan, *Ararat: The Shooting Script* (New York: New Market Press, 2002), 29.

[17] *Ararat,* 1:42:51–1:43:06.

[18] Hamid Naficy, "The Accented Style of the Independent Transnational Cinema: A Conversation with Atom Egoyan," in *Cultural Procedures in Perilous States: Editing Events, Documenting Change,* ed. George E. Marcus (Chicago: The University of Chicago Press, 1997), 219.

view of her relationship to an otherwise complex history. In grappling with the shortcomings of Ani and Rouben's ideas, Egoyan's *Ararat* suggests that both a formulaic poetic license and historical truth rely too heavily on static cultural cues and sources of identification.

The third scene, which should be read as a counterpoint to the previous two, involves Ani's son Raffi, who travels to historic Armenia to try to understand his people's history and his father's past. (The latter was killed because he attempted to assassinate a Turkish diplomat.) On his return to Toronto, Raffi is held at the airport by the customs official David upon suspicion of smuggling drugs in unexposable film cans. Raffi undergoes an extensive interrogation, during which he shows David video documentation of his trip. Raffi's video recording and voiceover can be read as an alternative to Saroyan's film, which casts itself as an historical drama. Whereas Saroyan's film relies upon recognizable heroes, villains, and tropes in order to facilitate automatic audience identification with its subject, Raffi's film embraces the complexity of its subject at the risk of estranging viewers. In contrast to the unambiguous placement of Mt. Ararat in Saroyan's film, Raffi presents the distanced mountain through the grainy lens of his camcorder, thereby heightening both his own and the audience's critical awareness of the site's *unfixed* representational potential. Similarly, if the scene where Marty scolds Ani's interruption of the film set exhibits Marty's confidence in his mission as an actor and the film's equally confident messaging, then Raffi's tone of uncertainty in his voiceover stands in direct opposition to Marty's attitude.[19] In terms of its content and acting, Raffi's film undercuts the moralizing bent in Saroyan's film. The audience listens to Raffi grapple with his identity based on a loose understanding of collective identity, one that involves a series of unanswered questions. The self-starring and self-made film is as much about Raffi's coming to terms with his identity as it is about his place in the narrative.

Raffi's voiceover, practically overwhelmed by a series of unscripted questions, suggests that the formation of identity relies on a conversational process. His questions attest to the inextricable link and inherent tension between personal and collective experience: "What am I supposed to feel when I look at these ruins? Is this proof of what happened? Am I supposed to feel anger? Can I ever feel the anger that Dad must have felt when he tried to kill that man? What's the legacy he's supposed to have given me? Why can't I take any comfort in his death?"[20] Raffi's train of thought pits the automatic anger that the ruins of the past would evoke for many Armenians against a contemplative attitude that strives to work through the connections among collective trauma, family history, and individual identity. The questions, then, should not be read for their answers—after all, the film does not offer any—but as a testament to the importance of actively positioning the self with respect to multiple narratives. Defying any type of materialist explanation, the open-ended questions represent identity construction as a fluid, ongoing series of highly personal negotiations between existing narratives and narratives in-progress. Although Raffi never expresses a fully realized sense of identity, the screenplay indicates that he returns from his

---

[19] Romney, 176.

[20] Egoyan, 63.

trip "transformed."²¹ The film only alludes to this transformation when Raffi recounts that he felt the presence of his father's ghost, after which "the meaning of things changed" for him.²² Some may argue that Raffi's newfound insight, with its lack of specificity, is entirely lacking in substance. Alternatively, the film's refusal to articulate the precise change in Raffi's thinking successfully emphasizes the agency of individuals in the making and *evolution* of collective identity.

Raffi's voiceover also illustrates the difficulties of representing the Catastrophe. As he puts it, "When I see these places, I realize how much we've lost. Not just the land and the lives, but the loss of any way to remember it. There is nothing here to prove that anything ever happened."²³ Here Raffi confronts the difficulty of representing the Event when material visibility has faded. Whereas his mother Ani treats the Genocide as an object of knowledge, Raffi considers it a source of evolving meaning. Ani's approach closely resembles what LaCapra has described as the process by which the "founding trauma" of genocide becomes "the valorized or intensely cathected basis of identity."²⁴ The film's challenge to this approach is conveyed through Raffi, for whom the Catastrophe poses more questions than answers about identity. Through Raffi's alternative film, Egoyan suggests that because the trauma of the Catastrophe can never be fully known, it unsettles the identity of surviving generations.

Raffi's character, then, uncovers the flaws in the static, absolute sense of identity upheld by the other Armenian diasporans in the film. He complicates the understanding of the self as it relates to cultural and personal history and the history of perceived Others. It is therefore appropriate that he spends most of his time at an airport, the ultimate liminal space, where diverse cultural currents and national identities intersect. Through Raffi's characterization, *Ararat* reimagines the definitions of diaspora, which despite their focus on transnational movement, traditionally emphasize the importance of a point of ethnic origin for group identity and consequently ignore historic, cultural, and experiential commonalities shared across ethnicities.²⁵ Raffi is the only character who rejects an exclusionary approach to diaspora by giving voice to the non-Armenian transnationals in the film, namely the half-Turkish actor Ali and Raffi's stepsister Celia, who are marginalized by the other characters in the film. For example, while Saroyan insults Ali and refuses to talk to him about personal matters, Raffi engages the actor in a conversation about his perspective on playing the role of "the evil Turk" in Saroyan's film.²⁶ Just as Saroyan shuts Ali out, Ani hinders her stepdaughter Celia's attempts to understand the cause of her father's death, refusing to tell her about the details of their last interaction. Moreover, Ani treats Celia's father's death as frivolous

---

[21] Ibid., 99.

[22] *Ararat*, 1:48:20-1:48:25.

[23] Egoyan, 63.

[24] Dominick LaCapra, *Writing History, Writing Trauma* (Baltimore: The Johns Hopkins University Press, 2001), 23.

[25] Floya Anthias, "Evaluating 'Diaspora': Beyond Ethnicity?" *Sociology* 32, no. 23 (1998): 558.

[26] *Ararat*, 55:03-1:04:53.

in comparison to the death of her first husband, who died for a national(ist) cause.[27] Raffi, on the other hand, empathizes with Celia for her loss, recognizing it as equivalent to his own experience of losing a father, and he makes multiple attempts to bridge the differences between stepmother and daughter throughout the film. The transversalism Raffi cultivates through these interactions allows *Ararat* to gesture toward a more inclusive approach to diaspora that seeks social solidarity across ethnic boundaries.[28]

Ararat presents its position on identity through its treatment of storytelling and representation. The film is concerned not with the definitive story of the Catastrophe, but rather with the crafting of traumatic narratives and their role in the construction of identity. This approach allows viewers to see identity as a process that relies on cultural production (the making of Saroyan's and Raffi's films, the production of Arshile Gorky's painting, the publication of Ani's book on Arshile Gorky). In one of her lectures on Gorky, Ani emphasizes the "expressive differences" between the painting *The Artist and His Mother* and the original photograph that inspired it: she cautions her audience not to read the painting as a painted re-depiction of the photograph.[29] Though the film repeatedly upends Ani's identity politics, it also arms her with the film's representative statement on aesthetic agency. Her insistence on the difference between the photograph and painting should be applied to the film's vision of identity and its relationship to history. Individuals engaged in narrating their stories transform transnational identities, which are inevitably the product of a negotiation between past and present.[30] *Ararat*'s refusal to restate the truth of the Catastrophe should be read as a gesture that empowers individuals to take part in producing narratives that would otherwise be at the mercy of those powerful enough to write history.

---

[27] Ibid., 35:30-37:10 and 1:15:00-1:16:35.

[28] Transversalism refers to the "process by which people designated as enemies or others form new understandings of each other [and] create spaces in which militarized ethno/national identities are examined and reworked." Lisa Botshon and Melinda Plastas, "Homeland In/Security: A Discussion and Workshop on Teaching Marjane Satrapi's *Persepolis*," *Feminist Teacher* 20, no. 1 (2009): 2-3.

[29] *Ararat*, 18:57-20:00.

[30] Along these lines, Stuart Hall has argued that "[Cultural identity] is a matter of becoming as well as being. It belongs to the future as much as to the past. It is not something which already exists, transcending place, time history and culture. Cultural identities come from somewhere, have histories. But, like everything which is historical, they undergo constant transformation. Far from being eternally fixed in some essentialized past, they are subject to the continuous play of history, culture and power." Stuart Hall, "Cultural Identity and Diaspora," in *Theorizing Diaspora*, ed. Jana Evans Braziel and Anita Mannur (Oxford: Blackwell Publishing Ltd., 2003), 236.

THE CHANGING ROLE OF LANGUAGE IN THE CONSTRUCTION OF ARMENIAN IDENTITY AMONG THE (AMERICAN) DIASPORA

Shushan Karapetian

This article examines the evolution of language ideologies and various forms of engagement with a heritage language in a diasporic setting. More specifically, it focuses on the role that language has played in the conceptualization of "Armenianness" among successive generations of Armenian-Americans. Tracing the shift of language from an implicit vernacular to a deliberate validation of heritage, in which it is perceived as both threatened and valued, this article aims to underscore the divergence between the reduced use of the language as vernacular and the increased symbolic engagement with the language. In other words, although the Armenian language's primary function and instrumental value as a vehicle for communication has narrowed in scope, the symbolic value invested in the language has expanded. Employing Jeffrey Shandler's framework of *postvernacularity*,[1] defined as a unique mode of engagement with the language in which the secondary level of a language's symbolic value is privileged over its principal role as a medium of communication, a special relationship with Armenian will be depicted, in which socialization into an affective ideology is prioritized over socialization into actual language use.

First, it is important to clarify and expand some of the terminology that will be used throughout the article. As formulated by Elinor Ochs, language socialization is concerned with two aspects of human behavior: first, it investigates how novices are socialized to use language, and second, it explores how language is used to socialize novices to become

---

[1] Jeffrey Shandler, *Adventures in Yiddishland: Postvernacular Language and Culture* (Berkeley and Los Angeles: University of California Press, 2008).

effective and competent members of their society.² Language socialization concentrates on the language used by and to novices, and the relations between this language use and the larger cultural contexts of communication – local theories and epistemologies concerning social order, local ideologies and practices concerning socializing novices, relationships between the novice and the expert, and so forth.³

Moving on to language ideology, Judith Irvine defines it as "the cultural (and sub-cultural) system of ideas about social and linguistic relationships, together with their loading of moral and political interests."⁴ Language ideologies act as interpretive links between a sociocultural context and linguistic forms and resources,⁵ shaping the understanding, evaluation, and deployment of these forms and resources from basic linguistic foci, such as vocabulary choice,⁶ to wider social and political foci, such as the hierarchical ordering of languages and dialects, within a community.⁷ In the Armenian case, the impact of dispersal and settlement in host countries, on the one hand, and the dominance of majority languages on the status of the Armenian language and the linguistic and attitudinal behavior of the Armenian community members, on the other hand, have been tremendous.⁸ Consequently, Armenian has continuously functioned in a bi- or multilingual environment,⁹ a factor that has undoubtedly shaped the language ideologies of the various communities. In the U.S., however, it is important to note that a dominant ideology of monolingualism is prevalent, coupled with the fact that bi- and multilingualism have traditionally been looked down upon.¹⁰

Finally, the inherent link between language and identity must be highlighted. Indeed, in addition to serving as a means of communication, ideologies of language connect the

---

² Elinor Ochs, "Indexicality and Socialization," in *Cultural Psychology: Essays on Comparative Human Development*, eds. James W. Stigler, Richard A. Schweder, and Gilbert Herdt (Cambridge: Cambridge University Press, 1990), 287-308.

³ Agnes W. He, "Heritage Language Socialization," in *Handbook of Language Socialization*, edited by Alessandro Duranti, Elinor Ochs, and Bambi B. Schieffelin (Oxford: Blackwell Publishing, 2011), 587-609.

⁴ Judith T. Irvine, "When Talk Isn't Cheap: Language and Political Economy," *American Ethnologist* 16 (1989): 255.

⁵ Paul V. Kroskrity, "Language Ideologies," in *A Companion to Linguistic Anthropology*, ed. Alessandro Duranti (Malden: Blackwell Publishers, 2004), 496-517.

⁶ Michael Silverstein, "The Uses and Utility of Ideology: A Commentary," in *Language Ideologies: Practice and Theory*, eds. Bambi B. Schieffelin, Kathryn A. Woolard, and Paul V. Kroskrity (Oxford: Oxford University Press, 1998), 123-45.

⁷ Rosina Lippi-Green, *English With an Accent: Language, Ideology, and Discrimination in the United States* (New York: Routledge, 1997).

⁸ Arda Jebejian, "Western Armenian is Nearing Extinction," *The Armenian Mirror-Spectator*, April 13, 2012.

⁹ Peter S. Cowe, "Amēn tel-hay kay: Armenian as a Pluricentric Language," in *Pluricentric Languages: Differing Norms in Differing Nations*, ed. Michael Clyne (Berlin: Mouton de Gruyter, 1992), 325-45.

¹⁰ Netta Avineri, "Heritage Language Socialization Practices in Secular Yiddish Educational Contexts: The Creation of a Metalinguistic Community" (PhD diss., University of California, Los Angeles, 2012).

language in question with the identity of a particular group or speaker.[11] In discussing the connections between language and identity, researchers in the areas of second language acquisition, language studies, and heritage language education highlight that identity is dynamic and socially constructed,[12] as well as negotiated in discourse and thus influenced by language, which serves as the medium for its negotiation.[13] Rather than a static category of possession,[14] identity is a process of constant negotiation, production, and performance.[15]

Identities are often linguistically constructed and negotiated because the connection between identity and language is "an intimate and mutually constructive relation,"[16] especially since language has important symbolic value[17] and plays a crucial role in establishing one's place and role in society.[18] Researchers view language not only as the medium of identity negotiation, but also as the source of identity interpretation.[19] Speakers of more than one language, including heritage language speakers, navigate within and among different language communities and thus negotiate their own identities in connection to these different languages and their power relations and social distributions in society. Thus, "the

---

[11] Paul V. Kroskrity, "Language Ideologies," in *A Companion to Linguistic Anthropology*, ed. Alessandro Duranti (Malden: Blackwell Publishers, 2004), 496-517; Lippi-Green, *English With an Accent*; Silverstein, "The Uses and Utility of Ideology."

[12] Mariana Achugar, "Writers on the Borderlands: Constructing a Bilingual Identity in Southwest Texas," *Journal of Language, Identity, and Education* 5 (2006): 97-122; Bonny Norton, *Identity and Language Learning: Gender, Ethnicity and Educational Change* (Essex: Pearson Education Limited, 2000); Guadalupe Valdés, "Heritage Language Students: Profiles and Possibilities," in *Heritage Languages in America: Preserving a National Resource*, eds. Joy Kreeft Peyton, Donald A. Ranard, and Scott McGinnis (Washington McHenry: Center for Applied Linguistics and Delta Systems, 2001), 37-77; Kendra R. Wallace, "Situating Multiethnic Identity: Contributions of Discourse Theory to the Study of Mixed Heritage Students," *Journal of Language, Identity, and Education* 3 (2004):195-213.

[13] Julie A. Beltz, "Second Language Play as a Representation of the Multicompetent Self in Foreign Language Study," *Journal of Language, Identity, and Education* 1 (2002): 13-39; Cristina Ros i Solé, "Autobiographical Accounts of L2 Identity Construction in Chicano Literature," *Language and Intercultural Communication* 4 (2004): 229-41; M. Warschauer, "Language, Identity, and the Internet," in *Race in Cyberspace*, eds. Beth Kolko, Lisa Nakamura, and Gilbert Rodman (New York: Routledge, 2000), 151-70.

[14] Adriana Val and Polina Vinogradova, "What is the Identity of a Heritage Language Speaker?," *Heritage Briefs* (2010). Center for Applied Linguistics. http://www.cal.org/heritage/pdfs/briefs/what-is-the-identity-of-a-heritage-language-speaker.pdf

[15] Robert Crawshaw, Beth Callen, and Karin Tusting, "Attesting the Self: Narration and Identity Change During Periods of Residence Abroad," *Language and Intercultural Communication* 1 (2001): 101-19.

[16] Beltz, 16.

[17] Li Wei, "Dimensions of Bilingualism," in *The Bilingualism Reader*, ed. Li Wei (New York: Routledge, 2000), 3-25.

[18] Pauline G. Djité, "Shifts in Linguistic Identities in a Global World," *Language Problems and Language Planning* 30 (2006): 1-20.

[19] Warschauer, "Language, Identity, and the Internet."

identity of heritage language speakers is co-constructed and contextualized as they maintain and build connections with both or multiple languages and cultures."[20]

Sociologists and linguists Anny Bakalian and Bert Vaux have both observed and presented in broad strokes the shift in the role that language assumes in the conceptualization of "Armenianness" among various generations of Armenian-Americans. Bakalian and Vaux categorize Armenian-Americans speakers according to two categories. On one side are the foreign born, older, and highly proficient Armenians, for whom,

> The link between language and ethnicity is vital – it is the essence of identity, authenticity, and uniqueness. In this line of thinking, the particular structural characteristics of Armenian are believed to cause, lead, force, constrain, and require its speakers to know, do, intuit, appreciate, and resonate the way they do. Armenian is viewed and experienced as a dynamo that generates sensitivities, skills, abilities, and understanding unique to its community of speakers.[21]

In opposition to this traditional "Armenianness," the other side, consisting of second and third generation Armenian-Americans, espouses a more symbolic "Armenianness," one that is not dependent on actual knowledge or practice of language.[22] Generation is the most significant variable in terms of where Armenian-Americans stand on the continuum between the traditional and symbolic definition of "Armenianness," as well as their language retention. The longer the generational presence, the lower the proportion of people of Armenian descent who can speak, read, or write Armenian and who believe that one must speak Armenian to be Armenian.

What Bakalian and Vaux do not provide, however, is an explanation and analysis of how this journey, from language as intrinsic, to language as symbolic, occurs. Here, a microscopic snapshot of the early stages of this transformation will be presented through the analysis of interviews with first and a half and second-generation heritage language speakers, demonstrating how actual proficiency *in* the language falls victim to elevated ideologies *about* the language. In the Armenian diasporic context, heritage speaker describes someone who is raised in a home where Armenian is spoken, who speaks or, at least, understands the language and who is, to some degree, bilingual in Armenian and English.[23] The subjects in this study include college age youth who had immigrated to the U.S. at a very young age or

---

[20] Val and Vinogradova, 3.

[21] Bert Vaux, "The Fate of the Armenian Language in the United States," Paper presented at the Armenians of New England Conference, Harvard University, Cambridge, Massachusetts, April 10, 1999, http://www.academia.edu/3881620/The_Fate_of_the_Armenian_Language_in_the_United_States.

[22] Anny Bakalian, *Armenian-Americans: From Being to Feeling Armenian* (New Brunswick: Transaction Publishers, 1993).

[23] Guadalupe Valdés, "Heritage Language Students: Profiles and Possibilities," in *Heritage Languages in America: Preserving a National Resource,* eds. Joy Kreeft Peyton, Donald A. Ranard, and Scott McGinnis (Washington McHenry: Center for Applied Linguistics and Delta Systems, 2001), 37-77.

were the children of Armenian immigrants, with varying degrees of proficiency in Armenian.

For heritage speakers, several prominent affective elements emerged in connection with the significance and value of knowing the Armenian language, all indicating the undeniable link between language and identity and the integral function of language as a central vehicle of cultural preservation and transmission. A very strong degree of what renowned linguist Joshua Fishman labels positive ethnolinguistic consciousness was present, including a sense of sanctity, kinship, and moral imperative.[24] The fragile nature of Armenian existence in a diasporic setting with a small worldwide population recurrently stood out as the source for a need to claim and take ownership of their heritage and language. As a result, heritage speakers have internalized a moral responsibility for cultural preservation accompanied by a concurrent fear of loss of this heritage in light of the visible assimilation they witness around them. Often, in speakers' narratives, these moral obligations and fears are communicated and mediated through parents and grandparents, highlighting the active socialization process. Similar findings were present in Ani Yazedjian's study of the process of ethnic identity development for Armenian adolescents, in which important cultural markers such as the Genocide, the Diaspora, cultural preservation, and language were viewed as tools for cultural transmission employed by socializing agents in the community.[25]

The awareness of a dispersed Armenian existence and the internalized moral obligation of claiming and perpetuating the heritage culture and language lead to a central belief among Armenian heritage speakers that equates being Armenian with knowing Armenian. Similar to the traditional and ascribed characterization of "Armenianness" held by foreign-born, older, and highly proficient community members discussed above, first and a half and second-generation heritage speakers project a conscious understanding that Armenian identity is contingent on knowledge and practice of the language. This is often articulated as a compulsory equivalence in which claiming Armenian identity requires proficiency in Armenian.

Despite the strong sentiments outlined above, there are less overt and less obvious language behaviors and attitudes that conflict with the positive attitudes and public orientations towards Armenian, leading to reduced language use and lower proficiency. Essentially, Armenian is stripped of its utility and considered to be devoid of any practical or instrumental value. Knowledge of Armenian seems to bear no benefits outside of the emotional and personal realm, with no tangible material gains. Most importantly, when viewed as an obstacle to academic advancement (which is naturally centered on English), and as such, to the accompanying socioeconomic and social mobility, it is deprioritized and devalued. Often, the same parents and community members who have socialized their children into such elevated sentiments about the value of knowing Armenian are complicit

---

[24] Joshua Fishman, "What Do You Lose When You Lose Your Language?" in *Stabilizing Indigenous Language*, ed. Gina Cantoni, (Flagstaff: Center for Excellence in Education, Northern Arizona University, 2007), 71-81.

[25] Ani Yazedjian, "Learning to be Armenian: Understanding the Process of Ethnic Identity Development for Armenian Adolescents," *Journal of the Society for Armenian Studies* 17 (2008): 165-87.

in the lowering of its status in their day to day actions and behaviors.[26]

Possessing such a close association between language, ethnic identity, and the moral duty of transmission without the necessary linguistic proficiency to justify it can function as a double-edged sword, highlighting many of the inherent contradictions in such paradoxical formulations. The incongruities between the internalized elevated language ideologies and lower proficiency, as a result of negative attitudes and behaviors, lead to a state of cognitive dissonance,[27] involving multiple sources of tension and anxiety for heritage speakers. In psychology, cognitive dissonance describes the excessive mental stress experienced by an individual who holds two or more inconsistent beliefs, ideas, or values at the same time. This type of discomfort may also emerge within a person who holds a particular belief but performs a contradictory action. In her analysis of the responses in the National Heritage Language Survey, Kagan observes that because of their daily existence in two cultures, heritage language speakers may be affected by various types of cognitive dissonance.[28] The first kind, related to an inability to express themselves easily in the home language, often manifests in expressions of embarrassment and shame. These types of intense emotions stem from heritage speakers' conception of "the heritage language as an inherent part of [their] being ('*my* language'), and ... a clear marker of 'belonging uncertainty' as they cope with their inability to perform this 'belonging' properly."[29]

In the case of the heritage speakers interviewed for this study, they clearly subscribe to the "language-and-identity" ideology, which emphasizes the integral connection between a person and his/her native language.[30] However, their lack of high competence in the heritage language does not allow them to fully meet the requirements set by the ideology. Thus, they are keenly aware of the contradictions inherent in the fact that they define an Armenian person as someone who speaks Armenian *well*, and furthermore they self-identify as Armenian, yet for the most part they are English dominant. As a result of this inconsistency, there is a great tension in claiming full access to Armenian identity due to the lack of the required linguistic proficiency. Often, this is expressed in the negation of the equation, so that the consideration of someone who is Armenian but does not know Armenian well results in feelings of shame and embarrassment.

In addition to the tensions related to accessing and claiming Armenian identity, divergent language ideologies lead to a related source of anxiety connected with feelings of guilt and shame, not only for lacking proficiency in Armenian, but also for their subsequent inability

---

[26] Shushan Karapetian, "*How Do I Teach My Kids My Broken Armenian?*": *A Study of Eastern Armenian Heritage Language Speakers in Los Angeles*, Doctoral dissertation (The University of California, Los Angeles, 2014).

[27] Leon Festinger, *A Theory of Cognitive Dissonance* (Stanford: Stanford University Press, 1957).

[28] Olga Kagan, "Intercultural Competence of Heritage Language Learners: Motivation, Identity, Language Attitudes, and the Curriculum." *Proceedings of Intercultural Competence Conference* 2 (2012): 72-84.

[29] Ibid., 75.

[30] John Myhill, "Identity, Territoriality, and Minority Language Survival," *Journal of Multilingual and Multicultural Development* 20 (1999): 34.

to fulfill the moral responsibility of transmitting Armenian heritage through the language. Repeatedly and ubiquitously, heritage speakers expressed an inability to come to terms with the fact that they would be incapable of transmitting Armenian culture, history, language, and all the other features that comprise the Armenian heritage to future generations due to their lack of competence in Armenian. Although the questionnaire used for the interviews included a question toward the end about speakers' desire to teach their children Armenian, preceding questions related to the reasons for learning Armenian and/or the importance of knowing Armenian consistently elicited responses about the obligation to perpetuate Armenian culture across generations. Without exception, speakers raised the issue of transmitting their heritage to their children and the anxiety they experience at the possibility of failing to carry out this critically significant moral obligation.

Here the snapshot comes to an end, and we are left with the question of how these heritage speakers will deal with the stress of cognitive dissonance. Experts in psychology explain that people strive for internal consistency and are thus motivated to find a solution in order to alleviate the discomfort of dissonance. One option is to change the behavior, which, in this case, would be to acquire the needed proficiency in Armenian to meet the high demands of the ideology. We have already seen that this is too difficult of a task to take on and one that is not prioritized by socializing agents. The other alternative is to change the cognition, in this case to reconceptualize the boundaries of "Armenianness." Since the literal and metaphorical burden of Armenian as a moral obligation is too heavy to carry without the necessary linguistic proficiency, the sense of "Armenianness" and the linguistic capacity involved must be reformulated to reach some kind of internal peace. As a result, a new modus vivendi is worked out from which emerges a new relationship between language and culture, and language and identity, leading to a culture that may function with a sizeable constituency, independent of a vernacular speech community.

With the passing of generations and the predictable decline of linguistic proficiency that goes along with it, second- and third-generation Armenian-Americans have retreated to formulating a more flexible delineation of what it means to be Armenian. Qualities such as activism in the Armenian community and fighting for the Armenian cause, which mainly revolve around Genocide recognition and assistance to the Republic of Armenia, are highlighted and encouraged.[31] Furthermore, in the case of the Armenian community schools, for example, the visible failure to produce graduates who are proficient in Armenian has led to mission statements de-emphasizing linguistic competence and, instead, accentuating goals such as "developing a strong sense of national and spiritual values," providing an "Armenian upbringing," and instilling a motivation and inspiration "to be actively involved in the pursuit of the Armenian cause."[32]

---

[31] Hagop Gulludjian, "On the Demotion of Language as the Virtual Territory Sustaining Diaspora," paper presented at the Arpa 20th Anniversary Conference, Sherman Oaks, California, April 19, 2012.

[32] Shushan Karapetian, "'Where is the Line of Retreat?': Challenges Facing Armenian Schools in Southern California," paper presented at the Community Language Schools Conference, UCLA, Los Angeles, California, April 13, 2012.

As for language, it has ceased to be the self-evident foundation of a culture and the contemporary language of daily life. It has been stripped of its status as the organic, living, breathing, dynamic entity that generates a way of life, that functions as a vibrant avenue of discovering and experiencing the world, that transmits the intellectual and emotional repertoire of a people, that is at its core the main vehicle of communication, to become a lifeless but highly valued object that can voluntarily be accessed when needed and in dire need of preservation, similar to an inanimate cultural artifact one houses in a museum in order to admire or cherish it from afar. Language, on this side of the spectrum, has taken on a symbolic function, no longer as a means, but as an end to itself, a statement of one's Armenian heritage, one's devotion to things Armenian. It has become the utterance of a greeting here and there, the use of a few key words to punctuate speech, the piece of jewelry with one's first initial, or the framed poster of the alphabet, not as a guide for a child to practice his Armenian letters, but as a piece of Armenian art. Paradoxically, in this altered engagement with the language, the deeper the actual loss of the practical use of the language, the greater its symbolic value as a signifier of loss.

EFFECTS OF THE GENOCIDE, SECOND GENERATION VOICES

Rubina Peroomian

The children of survivors of the Armenian Genocide are affected by their parents' traumatic experience: regardless of how they perceived and treated the parents' traumatic past, regardless of how the survivors themselves dealt with that past. The psychological effects of the trauma impressed upon the family atmosphere, and familial relationship.

Armenian survivors, refugees, newcomers, an insignificant minority or ethnic group in the New World, made their humble nests, raised a family and worked hard to provide a healthy and happy environment for their children to grow free of the scars of the past they themselves carried. They did not always succeed. The family atmosphere, which most of these first-generation survivors unconsciously provided, was rigid and unhappy. Janine Altounian, a daughter of Armenian Genocide survivors, epitomizes her parental home as "un foyer ou la joie de vivre n'était guère à l'ordre du jour."[1] Nava Semel, daughter of Holocaust survivors, has a name for children who were born in such families. She calls them "Children of sad people," people who "forgot the habits of joy."[2]

The above testimony brings me to assert, at the outset, that transgenerational effect of a trauma is not a phenomenon unique to the Armenian Genocide, but a common trait in other genocides as well. Second generation Holocaust survivors, like Nava Semel and scholars, such as Alan Berger, Helen Epstein, and Eva Hoffman, among others, have dully examined this phenomenon.

Jack Danielian, psychologist, maintains, that "Trauma is contagious, and the contagion is likely to be insidious. All who come in contact with it can come away marked, including

---

[1] Janine Altounian, born in Paris, is an essayist, translator. Her book on her father's memoir is a valuable contribution to understanding the mindset of the descendants of Genocide survivors and their perception of the parents' past. See, Vahram et Janine Altounian, *Mémoires du génocide arménien. Héritage traumatique et travail analytique* (Paris: Presse Universitaire de France, 2009). The quotation above is from p. 118 of her essay, titled "Parcours d'un écrit de survivant," 113-47.

[2] Nava Semel, "Intersoul Flanking: Writing about the Holocaust," in *Second Generation Voices, Reflections by children of Holocaust Survivors & Perpetrators*, eds. Alan L. Berger and Naomi Berger (Syracuse: Syracuse University Press, 2001), 71.

victim, victim families and progeny..."³ This notion is further corroborated by Ervin Staub, genocide scholar. He believes that, "our identity is deeply rooted in racial, ethnic, or religious groups, members of the group who were not present still carry many of the burdens of the collective experience."⁴

The psychological effect is there, and the memory of the traumatic experience of the parent is transmitted, no doubt, but how deep and to what extent? The contributory factors gaging the intensity are many: first are the circumstances within the new, unfamiliar world in which the survivors took refuge: is the atmosphere favorable or prejudicial? How much hardship is involved in their everyday struggle for survival? Then there is the relationship within the family of survivors: did survivor parents keep silence and did not share their harrowing ordeal? Was the parent's life story a tightly kept family secret? Did the survivor parents incessantly talk about their harrowing experience and admonished their children not to forget? Was the generation born in these families attentive to and conscientious of the family history which was often unthinkably depressing and discordant with the new environment? Was this generation absolutely detached from the Old World, busy making a successful life in the New World and completely aloof vis-à-vis the parents' past experience or, on the contrary, was entangled in its grips? Finding answers to these and more questions, painfully complex, by examining the literature of second generation Armenian writers and poets as a reflection, an echo, a testimony, have become my preoccupation and the topic of my research and writing for the last decade or so.⁵

To reach an in-depth understanding of the second-generation survivor syndrome, I have first concentrated on the parent-children relationship as the source and the feeder inspiring or provoking the literary responses. In my categorization, classification, or typification, I first tested the responses of those children who were **denied entry into the parents' past world of darkness.**

Many survivors of the Armenian Genocide chose not to speak to their children about their past, mostly out of fear of the harmful effects that these stories of blood and death could produce in the children's juvenile psyche. They consciously or subconsciously protected their children from the paralyzing memory with which they themselves had to live. Then there was this inexplicable self-blame or a sense of shame and reproach for having survived while other members of the family met a torturous death and perished: a temporary conversion to Islam, a "service" to the gendarmes or the "rescuer" Muslim man, a rape, a circumcision. Whatever the dark secret of their survival, they chose not to reveal. Others futilely believed that by not speaking about their traumatic past, they will manage to forget it

---

³ Jack Danielian, "A Century of Silence, Terror and the Armenian Genocide," *The American Journal of Psychoanalysis* 70 (2010): 245-64.

⁴ Ervin Staub, "Healing and Reconciliation," in *Looking Backward, Moving Forward, Confronting the Armenian Genocide,* ed. Richard G. Hovannisian (New Brunswick and London: Transaction Publishers, 2003), 263-74. Quotation from p. 265.

⁵ The book encapsulating the results of my research is titled *The Armenian Genocide in Literature, The Second Generation Responds* (Yerevan: Armenian Genocide Museum–Institute, 2015).

and ward off the frequent incursions of painful memories and tormenting flashbacks. There were also survivors who, in their haste to start a new life in the New World and adopt new ways, refrained from revealing their past lest it set them apart from the mainstream society. And this is especially true in Armenian communities in Europe and North America, where discrimination, prejudice, and social rejection against the newcomers was the prevailing practice.[6] For still others, the burden of daily struggle for survival in a new and unfamiliar world weighed so heavily as to keep them from talking about their traumatic past. Within the family atmosphere of many survivors, therefore, speaking about the traumatic past, the Catastrophe was a taboo. It could only be hush-hushed in the narrow circle of family friends—mostly survivors themselves—or in brief references to it, often in the Turkish language, unintelligible to their children. In all cases, however, the shadow of the Genocide was ubiquitous. Mary Terzian, a daughter of Genocide survivors growing up in Cairo, Egypt, writes, "The effects of genocide were present in my mother's glassy eyes and my father's angry temper. It affected us all and will probably have its effect on a few more generations."[7]

Psychologists believe that keeping secrets in the family, especially of traumatic nature, potentially can damage the mental health of the family members and destroy relationships. It can even cause physical pain, depression and anxiety in the keeper of the secret, when the secret is internalized. Most vulnerable in this situation are the children growing up in that family atmosphere, feeling an unnamed anxiety about something that is being hidden from them. The anxiety grows into an affliction when they think of themselves as personally responsible for that mysterious something, something they cannot identify. Suzanne Handler, MED, author of *The Secrets They Kept*, identifies five effects of keeping secrets in the family: It can destroy relationships; can affect children's lives; can cause suspicion and resentment; can create a false sense of reality; can cause illness.[8] What would Handler think of the horrendous experience of the Genocide that many survivors tried to keep secret?

Virginia Haroutunian speaks of her mother's mysterious past in *Orphan in the Sands*. On different occasions, Virginia had questioned her mother, but she was adamant. "Don't ask, I don't want to talk about it."[9] Virginia resented her mother, that dry and cold woman who shrank with every word of endearment even if it came from her husband or children; who kept to herself, had no close friends, never visited her neighbors, refused to join any

---

[6] As David Kherdian writes of his mother, they were "victim[s] of America who/escaped the Turkish Genocide." See, David Kherdian, *On the Death of my Father and other poems* (Fresno: Giligia Press, 1970), from the poem "A Family of Four."

[7] Mary Terzian, *The Immigrants Daughter: A private battle to earn the right to self-actualization* (printed in the United States of America, 2005), xiv.

[8] Suzanne Handler, MED, "5 Reasons Why Keeping Family Secrets Could be Harmful," http://psychcentral.com/blog/archives/2013/08/22/5-reasons-why-keeping-family-secrets-could-be-harmful/

[9] Virginia Haroutunian, *Orphan in the Sands* (N.p. [Michigan]: S.p., 1995), 130.

Armenian groups.[10] She had even refused to dance at her own wedding.[11] Virginia pours out her frustration and the deprivations she endured growing up in Pontiac, Michigan. It was only towards the end of her mother's life, after she had gone through a series of painful treatments for her throat cancer, that she finally gave in and shared with her daughter her horrible story of death marches, the killings, the torture, the rape, family members separated and lost, the seven-year-old girl wandering alone, hungry, thirsty, and sick with diarrhea, and the Turkish homes she took refuge and then ran away from their maltreatment; and finally, the miserable conditions in the American orphanage in Kharbert. Virginia knew now the reason for her mother's strange behavior. The deep resentment was now turned into an irremediable compunction.

In *My Literary Profile: A Memoir,* Helene Pilibosian speaks of her parents, "their attitudes and expressions, their way of living in the Boston area and adjustment to this country, their stories and their anger, their relationships to one another."[12] They raised their children with the inflexible customs, outlooks, traditions, and set of values of the Old World, and molded their character and personality, even with some repression, only to find themselves in psychological conflict with their children as well as with the outside world. The dark shadow of the great loss they experienced influenced their behavior and the personality of the children they raised. Through Helene Pilibosian's narrative, comes across her mother's rigid personality and behavior. She was a shy, unsociable woman, an embarrassment when she despaired, and that happened every so often.[13] She was angry at her daughter for reasons Helene could not understand. It was saddening for Helene when, in a crowded place, her mother would insist on returning home immediately, "with a sudden attack of claustrophobia."[14] She was a strict mother, never showing affection. "Unfortunately, in our family there was little communication of tender emotions, the hug and the kiss seeming nonexistent and sometimes the positive feelings also."[15] She constantly criticized Helene and put her down. She considered her incapable of taking care of herself or doing housework.[16] She did not value her daughter's creativity and her success in poetry writing. Helene knew nothing about her mother's ordeal during the Genocide, and she admits that she never really showed interest in hearing her story until much later, as she studied that period of Armenian history. She interviewed her mother shortly before her death and was shocked and horrified hearing what she had been through. At the age of 91, Yeghsa had confessed about her being rescued by a Turk and having lived in a Turkish family. The confession made her so tense and distraught that she could not continue. Apparently, as her daughter had surmised, she

---

[10] Ibid., 42.

[11] Ibid., 25.

[12] Helene Pilibosian, *My Literary Profile: A Memoir* (Watertown: Ohan Press, 2010), 6.

[13] Ibid., 31.

[14] Ibid., 47.

[15] Ibid., 54.

[16] Ibid., 79.

had been raped, beaten, and left unconscious half-dead.[17] Now she understood the source of her mother's strange behavior, her mother with anxieties of surviving starvation, seeing her mother walk away never to return, seeing the dead and the dying, experiencing the insecurity of orphanages. She wished that her mother could have the opportunity to seek psychiatric help "to ease the burden of the hurt she carried to give her some satisfaction and tranquility in her life and render our lives with her easier and more satisfying."[18] And indeed, the family atmosphere pushed Helene into a **state of a severe depression**, incapacitating her, driving her to seek psychiatric help even shock treatment. And this became another subject of her mother's derision, suspicion, and criticism.

In *Passage to Ararat,* Michael J. Arlen remembers **hating his father, being afraid of him**, though at the same time he confessed he loved him. He admits that this hatred must have been because of the fear that he always felt "of being exposed in some way, or pulled down by the connection," connection with anything Armenian, Armenian and Turkish, the Turkish massacres of Armenians, and everything pejorative that came with it.[19] And at the end, in his later years, long after his father's death, when he came to the full realization of the Armenian past and his connection to it, he wondered why all this camouflage? Why did his father avoid the "collective unconscious" about his racial past which was his identity?

David Kherdian's father, Melkon, was a survivor of the 1909 Adana massacres:

My father came from Adana
but his father was born in Kharpert,
where, my mother says:
"Most of the people were educated
at the University," but
my father wasn't educated anywhere[20]

That is all he knew about his father's past, a past he could not, he would not want to penetrate. The father-and-son relationship was shrouded by untold but mysterious stories of torture and death creating an abyss between the two that lasted his father's lifetime. The result was an intricate feeling of **awe, respect, admiration**, and at the same time **fear, resentment**, and **alienation** that young David held for his father:

---

[17] Ibid., 21-2.

[18] Ibid., 170.

[19] Michael J. Arlen, *Passage to Ararat* (New York: Farrar, Straus & Giroux, 1975), 11-12.

[20] David Kherdian, "My Father," in *On the Death of my Father and Other Poems* (Fresno: Giligia Press, 1968, 1970), np. About the title poem of this collection "On the Death of my Father," William Saroyan has said "The title poem is one of the best lyric poems in American poetry" (http://www.davidkherdian.com/node/24). The poem is also included in *Forgotten Bread: First-Generation Armenian American Writers*, ed. David Kherdian (Berkeley: Heyday Books, 2007), 323.

legs like piano stumps
chest like a barrel (much
in the manner of Babe Ruth)
he growled like a crossed
lion when Armenian-angry,
but was bear silent when not

and died in a Milwaukee hospital
on my sister's fifteenth birthday
far from any home.[21]

In his "struggle for identity," David, born in Racine Wisconsin, **shunned away from everything Armenian and what his father represented**.[22] His "resistance to Armenian food," his "preference for everything American," his arguments with his father "over spading the / vegetable patch, painting the / garden fence ochre instead of blue,"[23] were all symptoms of an alienated soul caught between a past he despised and ignored and a world of the present around him he so enthusiastically embraced.

It is only after his father's death that Kherdian experienced the creeping into his heart of a new, reversed sensation which is not unusual or farfetched in these kinds of parent-children relationship. The posthumous bonding resulted in a feeling of guilt and regret. He began reading about the history of the Armenian massacres, and he found the explanation of his father's strange disposition.

Ellen Sarkisian Chesnut and her siblings grew up in San Francisco with their parents, both survivors of the Genocide. Looking back to her childhood, she attests that the shadow of Genocide dominated the family atmosphere. Her father never succeeded to cast off that shadow revealing its existence in his periodic outbursts of rage. Ellen's parents were anything but normal compared to the easygoing American families she knew. Her book, *Deli Sarkis: The Scars he Carried, A Daughter Confronts the Armenian Genocide and Tells Her Father's Story*, clearly shows the impact of the ordeal each one of her parents had experienced during the massacres and deportations.

> When we were growing up, his rages would come and go. Sometimes he would be so understanding and speak so softly and sweetly, like the father we always dreamed about. At other times he would be critical of us all and beside himself

---

[21] Kherdian, "My Father."

[22] For a broader biography of this author, see Aris Janigian's article on David Kherdian in *Forgotten Bread*, 318-21.

[23] Quotations from David Kherdian's "For My Father," in *Homage to Adana* (Fresno: Giligia Press, 1970), n.p.

with anger. I tried to understand, but could never discover the reason for this split-personality.[24]

The resulting reaction Ellen and her two sisters revealed was **alienation and distancing**. They sought to live away from their parental home and keep a safe distance from that depressing atmosphere. However, the solution did not provide them peace of mind and a chance to live a normal and unfettered life.

Overprotectiveness is often a characteristic trait in the family of survivors. The burden of losses they have experienced drives these parents to this harmful attitude toward their offspring, lest they too perish. Children grow up **insecure, exceedingly dependent on confirmations for their deeds to come from others, never able to make decisions about their life, their career, and their relationships**. Virginia Haroutunian's life is a case in point. Diana Der-Hovanessian explains this phenomenon as a post-genocide Diasporan-Armenian reality:

why Armenian children
are raised with so much wonder, as if
might disappear
at any moment[25]

Mary Terzian, in hindsight explains her father's overreaction to her walking down the busy streets of Cairo to the English Catholic School, in a time when political turmoil had just broken out in Egypt against British oppressors. "Of course, Father is a genocide survivor, one who has seen rape, murder, and death as regular casualties of political strife."[26] That is a big difference compared to the generation born in the Free World and without the tragic experience their parents have endured.

But the story must be told. Despite being shut off from the parents' past, or even if having learned about it all along, a realization after the parents' death or just before their death opens a complex world of suffering and survival of the whole nation for the son or daughter of the survivor. And the urge sets in: the story should be shared with the world. In some cases, the progeny has a chance to interview the parents before their passing, collect enough information, and concoct the life story of the survivor. In other cases, elderly parents are encouraged to write their memoirs. With the feeling of a filial duty or impressed by the macabre past of the parent and of the whole nation, or perhaps as a way of cleansing one's guilty conscious for not understanding or not wanting to understand the parent's dilemma in due time, he/she undertakes the task of introducing that story to make the world know

---

[24] Ellen Sarkisian Chesnut, *Deli Sarkis: The Scars he Carried, A Daughter Confronts the Armenian Genocide and Tells Her Father's Story* (Minneapolis: Two Harbors Press, 2014), 158.

[25] Diana Der Hovanessian, "Diaspora," in *About Time* (New York: Ashod Press, 1987), 22.

[26] Terzian, 184.

about that colossal crime, that is a crime against humanity committed with impunity.[27] In still other cases, the collected information or the rough draft of the parent's memoir—often times a late discovery—is used as raw material into a novel and introduced as the fictionalized story of the Armenian Genocide.

David Kherdian's *The Road from Home, The Story of an Armenian Girl* (1979), the story of David's mother, Veron Dumehjian, is one of the first examples and a debut for the author in this painful domain. Some thirty years later, in a flyer publicizing the book *The Forgotten Bread*, with seventeen writers and poets presented in the volume, a quotation from David Kherdian reads, "our stories contain us and reveal us and inform us, filling us with the real pride that comes from having lived and endured, not only with our lives intact, but with our stories told. For without our stories we are nothing." A later example is Agop J. Hacikyan's *A Summer Without Dawn* (2000). Driven to tell the story of his parents' and his people's suffering, Hacikyan is also in pursuit of the perpetuation of the memory of the Genocide, making sure that the memory will be transmitted from generation to generation.

Theodore Kharpertian had not cared to ask about his father's past all his life. Upon his mother's insistence, however, he took up the task of writing his story. He was forty-five and his father, Hagop, was already eighty-seven. Significant in *Hagop: An Armenian Genocide Survivor's Journey to Freedom* is Theodore's idea of asking his father's permission to be happy and to let go his guilty conscious of having so much, "stable, loving parents, a good home, a nation that, in principle, strove to guarantee my fundamental human rights."[28] All that in contrast with his father's pain of so much loss and injustice he suffered. Theodore's action is an evidence of subconsciously inherited sentiment which had a nevertheless shifted foundation. His father's feeling of guilt for having survived while others perished, very much common among the survivors of the Armenian Genocide, had gone through a transformation in the next generation's psyche, and the catalyst is the distance of time and place with the actual losses the first-generation survivors experienced.

Kay Mouradian's *A Gift in the Sunlight, An Armenian Story* (2005), is a testimony of a strange transformation in her mother's heart in her last years of life and her urging Kay to write and publish her story for the world to know. The book was a beginning for Mouradian's

---

[27] In Peroomian, *The Armenian Genocide in Literature: Perceptions of Those Who Lived through the Years of Calamity* (Yerevan: Armenian Genocide Museum–Institute, 2012, 2014), this author has introduced a number of memoirs by first-generation survivors published by their children, either in the original Armenian or in English or French translation. Examples are: *To Armenians with Love* (1996), Hovhannes Mugrditchian's memoir translated into English and published by his son, Paul Martin; *A Hair's Breadth from Death* [2003], Hampartzoum Chitjian's life story, dictated in Armenian, translated into English by his daughter Sara Zaruhy Chitjian, and published in Armenian and English editions; *Le journal de mon père* (2008), Hrant Sarian's diary translated into French and published electronically by his daughter, Louise Kiffer; and *Im housheri chanaparhov* (2009), Yervand Kyureghyan's manuscript published in its original form by his son Varouzh Kyureghyan with Introduction, explanations, and commentary.

[28] Kharpertian, Theodore, *Hagop: An Armenian Genocide Survivor's Journey to Freedom* (Belmont: Armenian Heritage Press, National Association for Armenian Studies and Research, 2003), 128.

dedicating her time for spreading the stories of the Armenian Genocide and to teach the youth about this unforgivable crime against humanity.

Ellen Sarkisian Chesnut's *Deli Sarkis: The Scars he Carried, A Daughter Confronts the Armenian Genocide and Tells Her Father's Story* (2014), works in the same path: the world should know what the Turks did which became the survivors' lifelong ordeal and their children's hardship to cope with it.

Research done on the children of Holocaust survivors indicates that, as Ervin Staub puts it, "the trauma of parents does not usually result in psychopathology in children. However, some of the children, especially those whose parents do not talk about their own suffering, are affected in their interpersonal relationships."[29]

In the new and totally unfamiliar environment Armenian refugees raised their children instilling in them a large dosage of Armenianness with admonishments like, "You are Armenian, and don't you forget." Or "My family perished in the desert and you have to compensate for their loss by staying close to your roots." Mary Terzian learned about the meaning of all this only when she was much older. Then she came to see,

> why graduation from Armenian school, even from kindergarten, is so meaningful in our immigrant community. We represent the sprouts of a massacred generation. Every commencement is regeneration from the ashes and triumph over the perpetrators who wanted to erase our nation from the face of the earth.[30]

The technique used was the accentuation of cultural, religious, and ethnic differences in favor of what is Armenian which often produced a vexing and discomforting stance vis-à-vis the mainstream society in which they eventually had to enter right out of **a strict Armenian home.** As a result, a sense of **alienation and marginality** was generated. Mary Terzian experienced this shock when she graduated from the Armenian Community School in Cairo and began to attend the Immaculate Conception School in town, run by Irish nuns. She observes,

> To be absorbed into the foreign culture means enduring transformation that is partly suicide, partly betrayal, and partly reincarnation. Which part of the identity will be kept, which sector will be reshaped, and which segment will be abandoned are subject to debate.[31]

---

[29] Ervin Staub, "Healing and Reconciliation," in *Looking Backward, Moving Forward: Confronting the Armenian Genocide*, ed. Richard G. Hovannisian (New Brunswick: Transaction, 2003), 265.

[30] Terzian, 61.

[31] Ibid., 182. It should be noted here that the Armenian community in Egypt, mostly in Cairo and Alexandria, had a long history of existence and prosperity with old established schools, churches, news media, cultural, and political organizations and institutions. Armenians maintained social and business interactions with mainstream Egyptian society. The refugees of 1915 deportations and massacres joined this established structure, but with an inclination to stay within the boundaries of their own community.

Vahé Oshagan pictures a case of total alienation from the family in a short story titled *Dektember 31* (December 31), where one day the son left the Armenian home with all its shackles and restrictions, went away and never looked back.

Gerard Chaliand was fully exposed to the stories of atrocity against his family. He was a son of a survivor growing up in France, with the stories of his grandmother who was also a survivor. Perhaps, it was the high dosage of this inculcated history of blood and tears that made him discard "the tracks of the tribe," distance himself from "the ring of women spilling their sorrow," forget the pain of "being the heir of a genocide... stuck in the throat as an aching shard." Chaliand's alienation lasted a quarter of a century, a period of experimentation with "adventures of all human beings," and of escalation in his highly respected career, to come back to the realization of the undeniable effects of the Genocide.[32]

Virginia Haroutunian speaks of her shock as she discovered that the lifestyle in her Armenian family, the everyday rituals and cultural practices were different form everybody else in her class. Her strict mother did not allow going out with friends or dating, which she considered "improper and immoral." "Only street girls sit in a car alone with a boy," she said.[33] Virginia was left with "a growing **sense of isolation and depression**." This harmful sense of isolation, generating an array of various consequences, in Virginia's case drove her towards overeating.

The polarization, as Kherdian attests, was confusing and destructive for this first generation born in America, or Europe, or any other discriminatory conditions in that matter. The result was sometimes "**bragging, lying, or withdrawal**,"[34] not normal in any case.

The duality between the Armenian home and the outside world in the Boston area was the source of constant tension in Helene Pilibosian's childhood resulting in fits of **resentment and rebellion at home and alienation and marginality outside**. "Home was very much Armenian, and school was very much American."[35] Children picked up this feeling of too much Armenianness from remarks by their teachers or other children. In schools or in workplaces Armenians were looked down especially if they spoke Armenian.[36] The result was a sense of humiliation, "*Armenian inferiority complex*"[37] spreading across the generation born to these survivors of the Armenian Genocide.

---

[32] Gérard Chaliand is an expert in armed-conflict studies and in international and strategic relations, but also a historian, writer, and poet. He was born in 1934 in France, in a family of survivors of the Genocide. I have discussed in more detail Chaliand's perception of his inherited past in *And Those Who Continued Living in Turkey after 1915: The Metamorphosis of the Post-Genocide Armenian Identity as Reflected in Artistic Literature* (Yerevan: Armenian Genocide Museum–Institute, 2008, 2012), 165-68.

[33] Haroutunian, 74.

[34] Kherdian, "Our Block," in *Forgotten Bread*, 328-330.

[35] Pilibosian, 37.

[36] Ibid., 40.

[37] Ibid., 39.

No matter what the perception of the family atmosphere, the second-generation survivor feels **the obligation to remember. The burden of remembrance is passed on.** This is demonstrated by the many ways which the memory is treated. Some take a pilgrimage to Western Armenia to find the town, the village, the house, their parents called home. They are fulfilling their parents' often times unspoken wish to find their home and follow the route they walked to say a prayer for the unburied dead. The artist within that Genocide survivor's descendant is inspired to dedicate a literary work to that unique experience. The outcome is more than often a well-researched and emotionally charged piece, a story enveloping the parents' life-long ordeal and his/her own affliction on their side.

The trouble and the pain in the process pays off at the end even if it takes great talent and perseverance in digging deep to find the big picture, chasing the silences or trying to fill the mysterious gaps in the parents' narratives. Alice Tashjian did just that to concoct her mother's story, *Silences: My Mother's Will to Survive*. She had realized that there were certainly gaps in her story: "Her stories were selected remembrances of events she chose to tell."[38] She was determined to complete the story and find out more clearly the reason behind her mother's abstinence and selective narrative. Margaret Ajemian Ahnert's *The Knock at the Door, A journey Through the Darkness of the Armenian Genocide* is another example of how the parent's memory is treated.[39]

The burden of remembrance is passed on inspiring literary works. Diana Der Hovanessian, vowed to remember and pass on the memory when she was only twelve and her father took her to Michigan to visit the grave of her grandmother Zarif.

> Standing at her grave I wanted
> to promise her something
> about remembering.
>
> I wanted to tell her
> if I had children they would know
> how she lost her arm. I wanted
> to tell her that although
> most of her family was killed by
> Turks those of us who were left
> were hers too
> but I didn't know
> how to talk to the dead.

---

[38] Alice Agnes Tashjian, *Silences: My Mother's Will to Survive* (Princeton, NJ.: Blue Pansy, 1995), 7-8.

[39] Margaret Ajemian Ahnert, *The Knock at the Door: A Journey through the Darkness of the Armenian Genocide* (New York: Beaufort Books, 2007). A thematic analysis of Ester/Gezeer Kateejeh's experience as an Armenian girl abused and exploited during the Genocide appeared in, Peroomian, *And Those Who Continued Living in Turkey after 1915*.

When I came home
I wrote my first poem.⁴⁰

Helene Pilibosian turned a cold shoulder toward her parents' and her people's past, did not show interest to hear their stories, but as she grew more secure and self-confident she dug into the history of her nation, and in her later years, with the powerful tool of poetic expression in her hand, after her parents were already gone, she set out to celebrate their memory and eternalize the history of her nation in the context of American present. Her life is a quest for peace and equilibrium in an existence that rose from **a chaotic mix of two worlds**, always in conflict with each other, contradicting each other and refusing to reach a synthesis.⁴¹ In her poetry reverberates her struggle to define her identity in scattered lines such as "the Turkish sword"; "the hoary history"; "The apricot spoke Armenian,/and there I found my seed"; "My mouth is dry/with stories of the desert/of Der-el-Zor so long ago"; "Remembrance is the epitaph/for the ghosts of humble glory."

In some second-generation survivors, the pain of that transmitted memory of the Genocide is so vivid and life-affecting as if **the past is lived like the present**. In the poem "An Inconvenient Genocide," Alicia Ghiragossian speaks of that ever-present affliction,

> Pain is always fresh
> hidden
> in the mysterious wiring
> of the billions of cells
> playing trick on our mind
> mingling with the vibrations
> of our blood
> infecting and affecting life.⁴²

The past is lived like the present, and sometimes it overshadows the present with such intensity that a slight association in the present life with that of the past takes the second-generation survivor to the place and the time in which the adopted past event has occurred. And this is despite the fact that he/she is not only a generation removed and thus away temporally, but also away geographically.

Diana Der-Hovanessian experiences this strange feeling of being transported in the past while driving in the dark, looking for street signs. A bizarre psychological turnabout generates a state of mind to see the sign "Channing" as *Chankeri*. Then on the next turn it is *Ayash* instead of "Ayer." And she is with Siamanto, Rouben Sevak, and Grigor Zohrab, and other Armenian poets and writers and intellectuals arrested in Constantinople on the eve of

---

⁴⁰ Diana Der Hovanessian, "How to Talk with the Dead," *RAFT* 11 (1997): 20-1.

⁴¹ See my review of Pilibosian's *History's Twist: The Armenians*, in the *Journal of the Society of Armenian Studies* 18, no. 2 (2009): 145-48.

⁴² Alicia Ghiragossian, "An Inconvenient Genocide," (Glendale: "A New Age" Publications, 2009), 14.

April 24 (11), 1915 and murdered in the *Ayash* prison and on the road from *Chankeri*. Lorne Shirinian writes about her and her close association with the dead land, Armenia. "In the time warp called America/you mythologize the dead land/find yourself stranded between fable and reality." And indeed, this close association produces hallucinatory encounters in her life. Shirinian continues,

> you spot Varoujan dancing in Harvard Square
> you bump into Siamanto at Wordsworth's bookstore
> you receive a phone call; Tourian is lost in Watertown.[43]

With that same type of association, Hakop Karapents is transported to Urfa. He is attending the opening session of the Urban Renewal Federation of America's (URFA) annual conference in Hartford.[44]

Living the years of calamity all over again, some survivors penned their memoirs to show the world the magnitude of the Catastrophe. Armenian organizations struggled in the labyrinths of world politics to make the world know about the committed crime and help to recompense. But the world responded with indifference and conveniently accepted the Turkish side of the story. The **perpetrator's continued denial, and the indifference of the world community become sources of anger and frustration** in the second-generation survivors, and they responded: Some idealized and eternalized in their literature the daring acts of young Armenians, the freedom fighters, in their acts of violence against the deniers, descendants of the perpetrators. They stressed the fact that the purpose of these acts is less to avenge their ancestors than to shake up the world's indifference to the Cause and the continued suffering of Armenian survivors of the Genocide. Hakob Karapents' *Seneak tiv 842* (Room number 842), Vahé Oshagan's *Ahabekich* (Terrorist) and *Telefone, Lizboni nahataknerun* (The Telephone, to the Martyrs of Lisbon), Boghos Kupelian's *Pasport, vep merorya hay azatamartikneru kyanken* (Passport, a novel about the life of Armenian freedom-fighters of our days) are examples. Others combatted the denialism giving birth to exquisite pieces of poetry. Hrand Markarian's *Mets Tarapank/The Great Agony,* Diana Der-Hovanessian's poems "Keri's Curse," "The Political Poem," and "At Mt. Auburn Cemetery," Alicia Ghiragossian's, "An Inconvenient Genocide" encapsulate the rage and frustration caused by callous denialism.

But the Turks continue to deny, and this means dying "every time/we are invalidated/through denial," Alicia Ghiragossian writes,

---

[43] Lorne Shirinian, "Armenian Poets, for DDH" in *Earthquake, Poems* (Lewistone, Queenstone, Lampeter: The Edwin Mellen Press, 1991), 18-9. Diana Der Hovanessian admired Daniel Varoujan (Varuzhan), a great poet who was murdered in 1915 as well. Bedros Tourian (Petros Dourian), was a Constantinople born Armenian poet, who in his short life (1851-1872) left a rich legacy of romantic poetry and drama.

[44] Hakob Karapents, *Mer nakhnikneri stvernere* [The shadows of our ancestors] in *Ankatar* [Incomplete], a collection of short stories (New York: Vosketar, 1987), 33-49.

So we
the children of survivors
of an inconvenient genocide
are reporting
with earthshaking resonance:

Yes. It happened.
And we have not forgotten.

It was the same rage and frustration that drove Mae Derdarian to take up a project she had overlooked for years. When she read an advertisement sponsored by the Turkish government stating that what happened was not a genocide, as Armenians claim, but a civil war between Turks and Armenians, she was outraged. She remembered a manuscript that her mother's best friend, Vergeen Meghrouni, had entrusted to her twenty years earlier, before she died in 1975, asking her to prepare it for publication. The manuscript contained her life story, her experience during the trying years of the Genocide and her survival. Derdarian set to work, and the result was *Vergeen: A Survivor of the Armenian Genocide*,[45] a book that received deserved publicity and accolades. Some reviewers regarded the book as the Armenian version of Anne Frank's Diary.[46] The book was Derdarian's powerful response to Turkish denial in which reverberates Vergeen's objective in writing her life story, "to immerse the reader in her story and to refute historical revisionists who deny and distort the facts of the Armenian holocaust." Vergeen had begun her narrative affirming "I was THERE! I was an EYEWITNES! I was a VICTIM!"[47] And what Vergeen had witnessed and endured is beyond imagination. She takes the reader to the killing fields, to listen to the cries of pain and the supplication to have mercy, to spare, "*Khntrem, khntrem*" (Please, please). And then the helpless moaning, "*Allah nerdehseen?*" (God, where are you?). Vergeen lost her innocence, her spirit and her faith to the atrocities, "My faith in religion was destroyed. I hadn't prayed or gone to church since fleeing from my Arab captors; and my aversion to prayer has continued since then."[48]

Building on Ervin Staub's concept of the process of the healing denied to Armenians, I conclude this cursory glance on the second-generation responses to the Armenian Genocide: historians are pushed to reiterate their irrefutable findings and probe deeper to discover new facts. Artists, writers, and poets use the medium best available to them to confront the Genocide in art, and prove in the way they know best, that what happened in 1915

---

[45] Mae M. Derdarian, *Vergeen: A Survivor of the Armenian Genocide* (Los Angeles: Atmus Press, 1997).

[46] In an important monograph, *Genocide, A Comprehensive Introduction* (New York and Abingdon: Rutledge, 2006), Adam Jones has chosen Mae Derdarian's book and Vergeen's experience as a case study to introduce the Armenian Genocide. The Chapter on the Armenian Genocide (101-23) houses this case study titled "One Woman's Story: Vergeen," 109-11.

[47] Derdarian, 1.

[48] Ibid., 213.

could not be less than genocide. The second-generation Armenian artistic expressions are inherent carriers of that state of mind and that mission. Their literature echoes the nation's collective psyche shaped by the violence, the pain of dispersion, the effects of self-accusation, the struggle to cope with a dual identity or a search for identity, the struggle to cast off the shadow of the past, and the effects of the past and present stance of the perpetrators and world bystanders. As Diana Der-Hovanessian writes,

> Even though your mother was a baby
> in Worcester, and safe
> and your father a young soldier
> in Murad's mountains
> and you a generation from being born,
> . . . .
> even without a single
> relative who lived to march,
> lived past the march. We are children of Der Zor.[49]

The literary representations of the Armenian Genocide will continue to shape the understanding of this unresolved injustice for generations to come. They will function as the most effective transmitters of memory, shoring up commitment to the national struggle. These literary representations will stand as a monument to the memory of the Armenian tragic past, but more than that, they will work as an impetus to find a way to resolve the tragedy in order to make national survival and perpetuation possible.

---

[49] Excerpt from Part 2 of the three-part poem, "Tryptich," titled "Why Sand Scorches Armenians," in *About Time*, 14.

List of Contributors
(in alphabetical order)

Gregory Areshian, a Foreign Member of the National Academy of Sciences of the Republic of Armenia, has had a long and distinguished academic career in the USA, Armenia, and the former Soviet Union. He is the author of more than 150 scholarly works published in five languages in twelve countries, mostly devoted to interdisciplinary studies in social sciences and the humanities with a special focus on the Middle East and Armenia in a broader historical context. They cover a chronological timespan from Prehistory to World War I. His last edited volume concerning the multidisciplinary study of empires was published by the University of California in 2013. The volume includes a chapter on the relationship between imperialism and the formation of Armenian identity. He serves as Professor of History and Archaeology in the College of the Humanities and Social Sciences at the American University of Armenia.

Sebouh David Aslanian is the Richard Hovannisian Endowed Chair of Modern Armenian History and Associate Professor in the Department of History at UCLA. Aslanian specializes in early modern world and Armenian history and is the author of *From the Indian Ocean to the Mediterranean: The Global Trade Networks of Armenian Merchants from New Julfa* (2011), which was the recipient of the PEN Center's Exceptional UC Press First Book Award and winner of the Houshang Pourshariati Iranian Studies Book Award. His essay "A Life Lived Across Continents: A Gobal Microhistory of Marcara Avachintz, an Armenian Agent of Colbert's *Compagnie des Indes Orientales*, 1666-1707" is scheduled to appear in *Annales: Histoire, Science Sociales* in 2018. Aslanian is now completing his second book manuscript dedicated to early modern global print history.

S. Peter Cowe is Narekatsi Professor of Armenian Studies at UCLA. Previously, he has held positions in Armenology at the Hebrew University of Jerusalem and Columbia

University, New York. His research interests include Late Antique and medieval Armenian intellectual history, the Armenian kingdom in the context of state formation across the medieval Mediterranean, and modern Armenian nationalism. The author of five books in the field and editor of nine, he is the past co-editor of the *Journal of the Society for Armenian Studies*. He has served on the executive board of the Society for Armenian Studies and Association Internationale des Etudes Arméniennes. A recipient of the Garbis Papazian award for Armenology, he was inducted into the Accademia Ambrosiana, Milan (Classe di Studi sul Vicino Oriente) in 2015.

Touraj Daryaee is the Maseeh Chair in Persian Studies and the Director of the Dr. Samuel M. Jordan Center for Persian Studies and Culture at the University of California, Irvine. He is interested in the history of late antique Iran and the ancient world. He is the author of *Sasanian Persia: The Rise and Fall of an Empire* (I.B. Tauris, 2009) and the editor of the *Oxford Handbook of Iranian History* (Oxford University Press, 2012).

Myrna Douzjian is a lecturer in the Slavic Languages and Literatures Department at UC Berkeley, where she teaches courses on Armenian language, literature, and film. Her articles examining the politics of twentieth-century Armenian literature have appeared in the *Brill Handbook of Oriental Studies* (2014) and the edited volume, *An Armenian Mediterranean: Words and Worlds in Motion* (2018). She is also the translator of contemporary Armenian poems and plays, which have been staged in the United States. Her current research focuses on post-Soviet literary culture as a paradigm for reexamining discourses on world literature.

Ani Honarchian is a doctoral candidate in Armenian Studies at the Near Eastern Languages and Cultures at UCLA. She finished her first masters in Translation Studies at the Tarbiat Moallem University in Tehran, Iran and her second masters in Iranian Studies with an emphasis on Armenian and Syriac hagiographical and historiographical account of the Sasanian era. She is interested in the interconnected history of the Late Antiquity and the development of early Christianity in the Near East.

Shushan Karapetian is a Lecturer in the department of Near Eastern Languages and Cultures at UCLA. Her dissertation, "'How Do I Teach My Kids My Broken Armenian?': A Study of Eastern Armenian Heritage Language Speakers in Los Angeles," won the Society for Armenian Studies Distinguished Dissertation Award in 2015. Her research interests focus on heritage languages and speakers, particularly on the case of Armenian heritage speakers in the Los Angeles community, about which she has written, presented, and lectured widely. Most recently, she was the recipient of the Russ Campbell Young Scholar Award at the Third International Conference on Heritage/Community Language in recognition for outstanding scholarship in heritage language research. She is currently serving on multiple committees both in the local Los Angeles and global diasporic Armenian communities aimed at reforming Armenian language instruction and promoting the use of the Armenian language.

Rubina Peroomian is an independent scholar who has lectured widely. She is the author of articles in English and Armenian in scholarly journals and has contributed chapters in books on Diasporan Armenian as well as genocide literature. She has also authored textbooks in Armenian on the Armenian Question for high school students and a manual (material, lesson plans, and suggested methods and strategy) on teaching the Armenian Genocide to K-9 students. Her publications in the field of genocide studies include *Literary Responses to Catastrophe: A Comparison of the Armenian and the Jewish Experience* (1993), *And those who Continued Living in Turkey after 1915, The Metamorphoses of the Post-Genocide Armenian Identity as Reflected in Artistic Literature* (2008 and 2012), *The Armenian Genocide in Literature, Perceptions of Those who Lived through the Years of Calamity* (2012, 2014), and *The Armenian Genocide in Literature, The Second Generation Responds* (2015).

Khodadad Rezakhani is an Associate Research Scholar of Late Antique history at Princeton University. His research concentrates in Central and West Asia in the first millennium CE. He is the author of *ReOrienting the Sasanians: East Iran in Late Antiquity* (Edinburgh, 2017) and the forthcoming *Creating the Silk Road: Travel, Trade, and Myth-Making* (I.B. Tauris, 2018).

Roman Smbatyan received his doctoral degree in the early modern history of Iran and the Caucasus from University of Isfahan, with a dissertation on "The Position of the Caucasus in the policies of Iran of Nadir's era (1734-1747)." He then worked as Assistant Professor in Oriental Studies Department of Yerevan State University, teaching the histories of Iran, the Caucasus, and Shi'ism. Upon moving to the United States, he worked as a visiting lecturer in the History Department at UC Irvine from 2014 to 2016. His research interests are on early modern Armenian-Iranian interactions, early modern Armenian local principalities, political and socio-cultural influences of Iranian civilization upon the Caucasian ethnic and religious groups, and state and identity formation processes in early modern and modern Armenia and the Caucasus. He has published five articles and co-authored a book.

Giusto Traina is professor of Roman history at the Sorbonne Université, Paris, and a senior member of the Institut Universitaire de France. He is the author of several books and articles. Formerly interested in ancient landscapes and techniques, he is currently involved in a long-term research project about ancient Armenia. His book *428 AD - An Ordinary Year at the End of the Roman Empire* (Princeton U.P., 2009), first published in Italy, is also translated into French, Greek, and Spanish. He was awarded the Cherasco Storia Prize for his book *La resa di Roma. Battaglia a Carre, 9 giugno 53 a.C.* (Laterza, 2010). During Spring 2017, he was Dumanian Visiting Professor at the University of Chicago. He was recently Fellow of the Bogliasco Foundation (February 2016) and the Topoi Excellence Cluster at Berlin (Spring 2016; Summer 2017), and DAAD senior fellow at the Freie Universität at Berlin (Summer 2018).

Printed by Libri Plureos GmbH in Hamburg, Germany